Hamlyn all-colour paperbacks

Charles Solomon

Mathematics

illu[illegible] by Kenneth Ody

Hamlyn · London
~~Paul Hamlyn · London~~
Sun Books · Melbourne

FOREWORD

This book sets out to dispel many common fears about the difficulties of mathematics. It firmly establishes at the outset just how dependent we are on our own ability to carry out all kinds of rapid mental calculations in everyday life – and in fact just how proficient we are. We tot up shopping bills, our score at cards, we cross roads in the rush hour in perfect safety – all with the help of mathematics. Once over this initial hurdle, the author explains in simple language some of the fundamental ideas and methods of arithmetic, algebra, geometry and trigonometry. The book, written primarily for the adult reader in need of encouragement, concludes with a special section of not-so-easy problems (and their solutions).

To my wife, who helped – as always

Published by The Hamlyn Publishing Group Ltd
London · New York · Sydney · Toronto
Hamlyn House, Feltham, Middlesex, England
In association with Sun Books Pty Ltd Melbourne.

Phototypeset by BAS Printers Limited, Wallop, Hampshire
Colour separations by Schwitter Limited, Zurich
Printed in England by Sir Joseph Causton & Sons Limited

CONTENTS

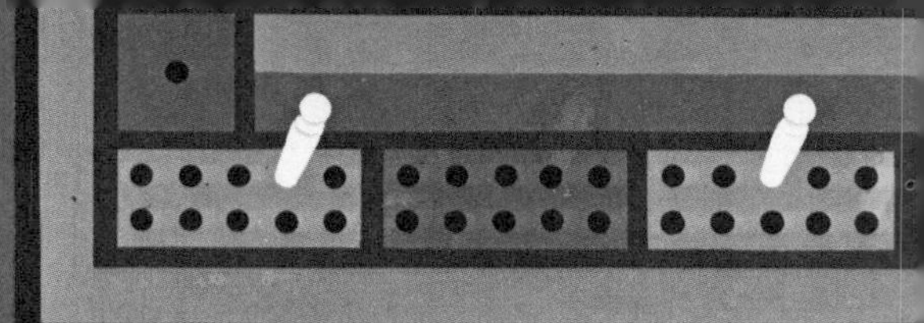

. . . fifteen-six,
two threes are
six, and
a pair's
fourteen . . .

MATHEMATICS IN EVERYDAY LIFE

'What would life be without arithmetic, but a scene of horrors?' – *Rev. Sidney Smith*

The game of cribbage calls for some quite nice calculations in the field of probabilities; and these are carried out more or less unconsciously by many players who hardly know the rudiments of mathematics. The academic player of two-handed cribbage who is dealt a five knows that, if he has not also been dealt a ten or a court card (which counts as ten) among his six cards, there is quite a good chance of a card of value ten being turned up in the pack. The actual chance is $\frac{16}{46}$ (nearly 3 to 1 against) compared with the only $\frac{4}{46}$ chance (more than 11 to 1 against) that a card of single nominated value, say a seven, will be turned up. On the other hand the non-mathematician probably has no idea of the odds, but he does not hesitate to retain a five.

When it comes to the show-down the mathematician will often feel abashed at the speed with which the other players reckon up their scores. As an example, suppose the player has in his hand the four and five of hearts and the six and queen of clubs and that the 'turn-up' card is the four of diamonds. The seasoned cribbage player will at once declare: 'Fifteen-six, two threes are six, and a pair's fourteen' while the tyro at the game (even though he be a mathematician) is still making up his mind that there are three ways of totalling 15 (queen and five, and six and five combined with each of the two fours), two straights of three cards each by combining each of the two fours with the five and the six, and a pair of fours. He will be hopelessly outpaced, although probably many of the players, if they were confronted with the written calculation $2 + (2 \times 2) + (2 \times 3) + 2$, would scratch their heads and plead that they are 'no good at ciphering'.

Using his own brand of elementary mathematics, a practised cribbage player can calculate his score almost without pausing to think and the speed of his calculations will often put to shame an opponent with a more academic approach to mathematics.

The first thing to remember about mathematics is that it is not simply a matter of writing down numerals, x's and y's; it is more a way of thinking. The man playing cribbage is, of course, using mathematics, but he does not call it by that name. He knows elementary mathematics all right, but he has not learned the language. And clearly you are not going to get very far with the study of Homer until you have familiarized yourself with the Greek alphabet.

Have you ever paused to consider the really quite difficult mathematical procedures you are using every day without realizing it? Think of the knowledge (though it be unconscious knowledge) of geometry that is required for a losing hazard at billiards. Or of the complicated calculations

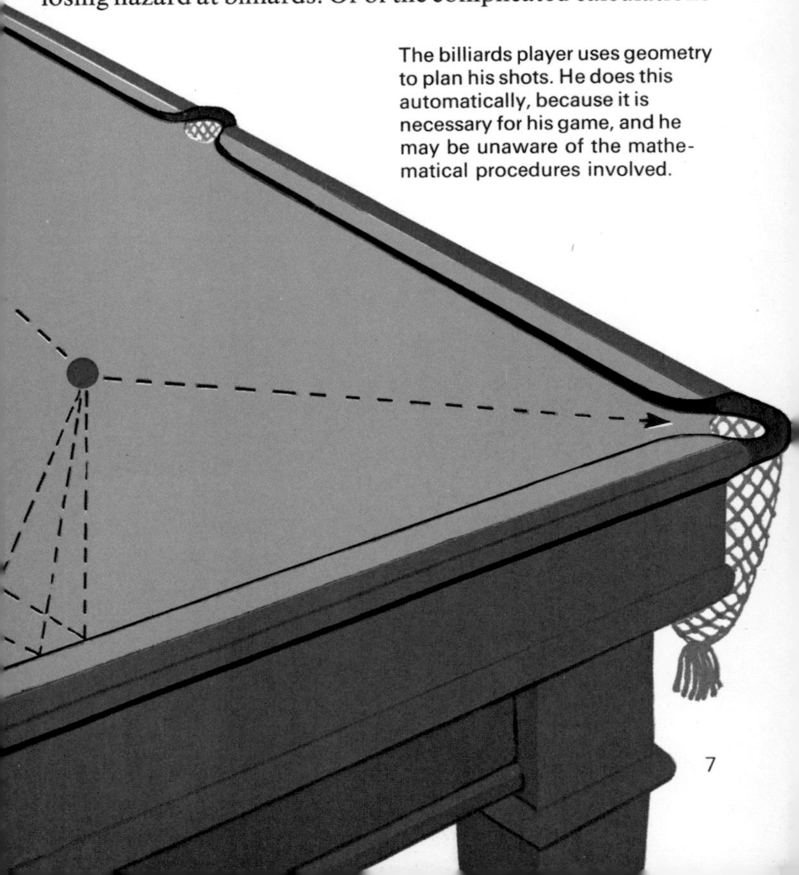

The billiards player uses geometry to plan his shots. He does this automatically, because it is necessary for his game, and he may be unaware of the mathematical procedures involved.

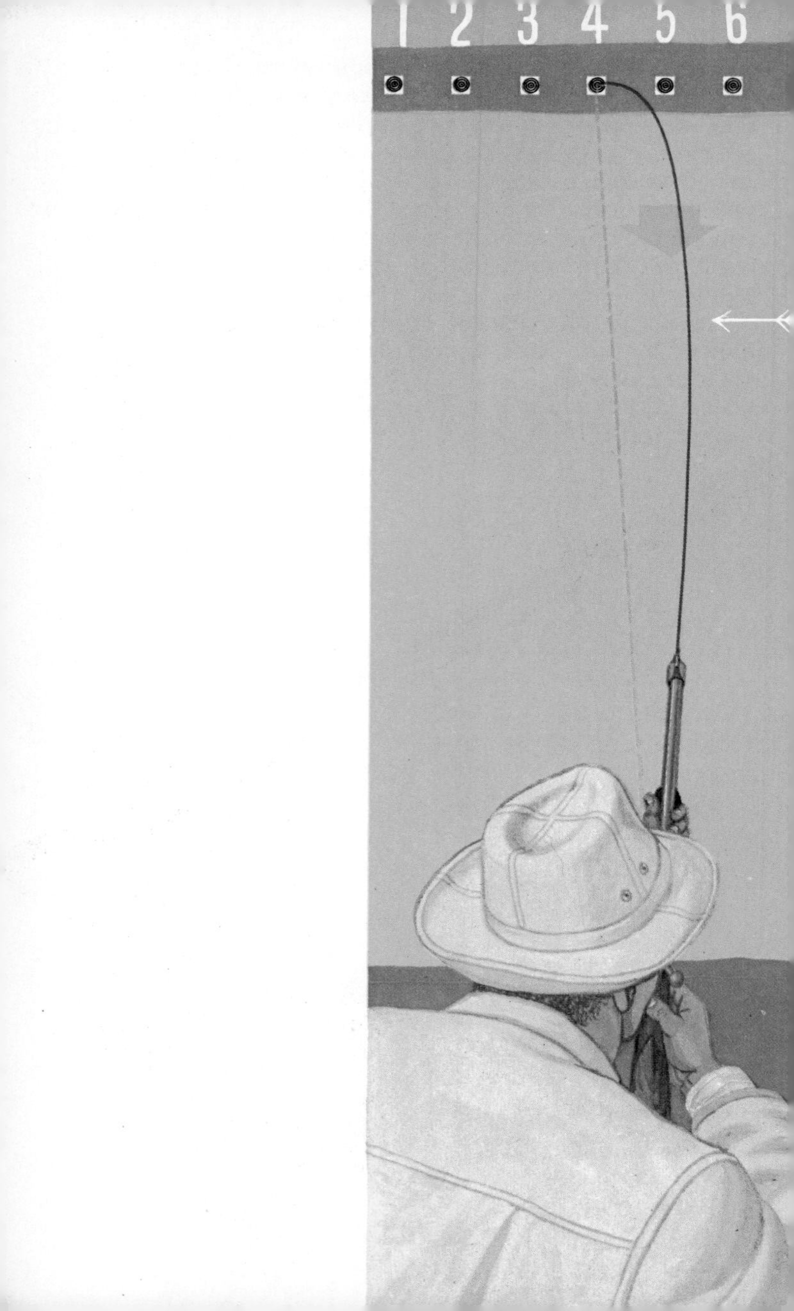
1 2 3 4 5 6

and estimates needed to hit an archery target. You must judge the size and distance of the target, the weight of the arrow as against the estimated velocity, and you must allow for cross-currents of air. Even if you could feed all this data into a computer it would do you little good; some of the conditions would have changed by the time you had lodged your information.

Again, a rifle-shooting competitor allows for wind and estimates the drop in the parabolic curve that the path of his bullet will take as the result of gravity. He does not call himself a mathematician and probably could not tell you what a parabola is; but if he isn't using mathematics what is he doing?

You have obviously from time to time crossed the road, and clearly, since you are reading these words, you have survived that perilous ordeal. The amount of mathematical calculation you have done (even though you may not be aware of it) might give pause to a Newton. You have estimated the speed of the cars approaching from

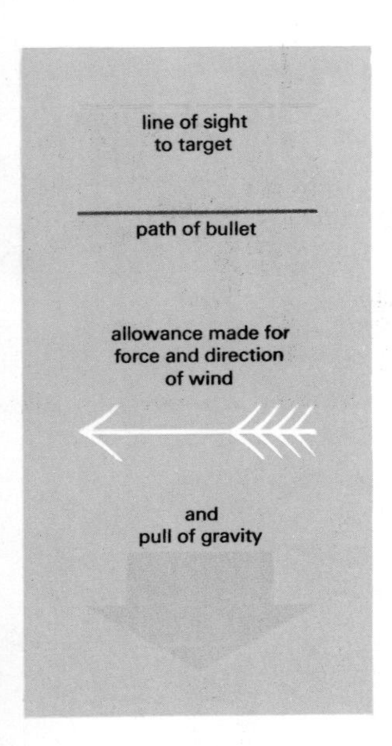

A rifle bullet travels in a curve (parabola) the shape of which is dictated by the forces of wind and gravity. The competitor may not know what a parabola is but unconsciously he uses mathematics each time he takes aim.

both directions as well as your own speed (allowing, I hope, a margin for safety) and you have decided the most direct route you can take while avoiding the traffic. Apart from making the estimates correctly, you will have used geometry, trigonometry and the calculus, even though you may deny with hand on heart even a nodding acquaintance with any of these subjects. And yet you accomplish this feat daily with repeated success.

How do you achieve this result? Simply enough really. It is because each one of us has tucked away in his skull a built-in computer. And it is a very reliable instrument indeed. It is as good (though of course not as fast) a computer as the latest model on the market; that is to say, it will never make a mistake unless you feed it the wrong

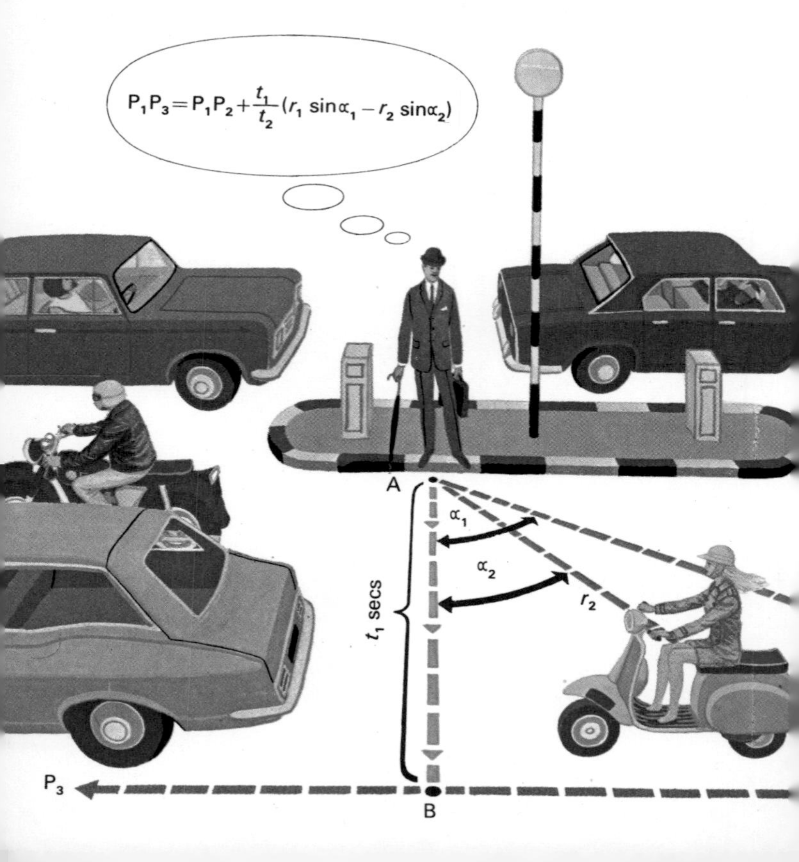

information. If you have misjudged the speed of the cars or the angle your crossing of the road should take, then you will have plenty of time to study mathematics in your hospital bed.

It is not the purpose of this book to 'teach' mathematics. A book twenty times its size would hardly be adequate for the purpose. We assume here that you learned some mathematics at school; that you have (like most of us) forgotten a good deal of it; but that you know enough of the subject to be able, say, to give change for a sum of money, to reckon what return you should get from an investment, to know something about such common geometrical forms as a triangle, a rectangle and a circle and recall something about their properties; and that you are not likely to faint on being confronted with such abstruse symbols as x, y^2 and $\sqrt{z}$. Provided you have that much mathematical technique (and, indeed, in these days of compulsory education it would be difficult *not* to have it) there is no reason why you should not be able to cope very comfortably with the contents of this book.

Before crossing the road it is a good idea to estimate the speed of the approaching traffic and your own speed, and then decide on the most direct route across to the other side. With this information in mind you then determine the best moment to make your crossing–a decision which involves the use of geometry, trigonometry and the calculus.

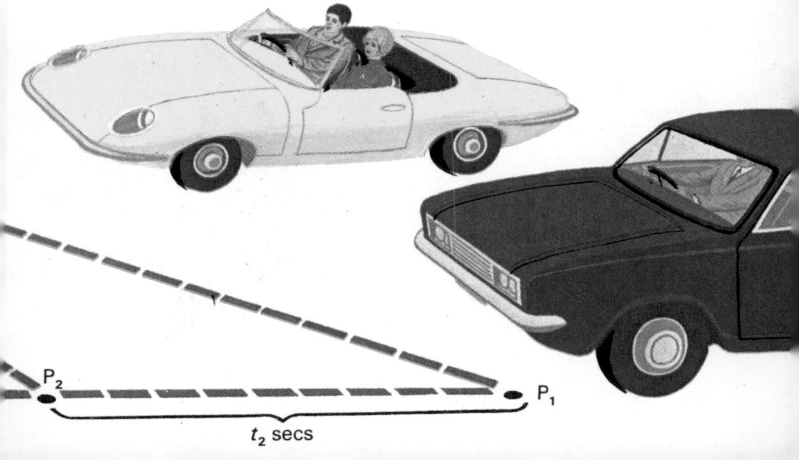

P_2

P_1

t_2 secs

Some of my younger readers may have had the good fortune to attend modern schools which teach the 'New Maths'. There is much to be said for this method of teaching provided (and this is a vital point) that the child starts it at an early enough age. It is undoubtedly a good thing to become acquainted with solid objects like spheres and cylinders before bothering about the way they are constructed, to recognize circles and ellipses before concerning yourself with the forbiddingly named 'conic sections'. But it is, as I have said, conditional on starting early enough. If you know sufficient of mathematics to have read and understood even this far, then I fear it is too late for you to revert to kindergarten methods, however admirable they may be; any more than you, an adult, can improve your knowledge of English by going back to 'the cat is on the mat'.

Also, there is a danger in the 'New Maths' that the pupil will neglect the duller chores that are inherent in and necessary to the subject. Playing with blocks may be useful as well as entertaining, but it is no substitute for learning your tables and acquiring some skill in manipulating numbers as distinct from shapes. The student who confines himself to the 'New Maths' is likely to finish up in the position of those who have learned a foreign language by well-publicized 'easy' methods. They may be able to converse fluently but know no syntax, which is all very well if you want to order an apéritif but not so good if you want to enjoy reading in the language.

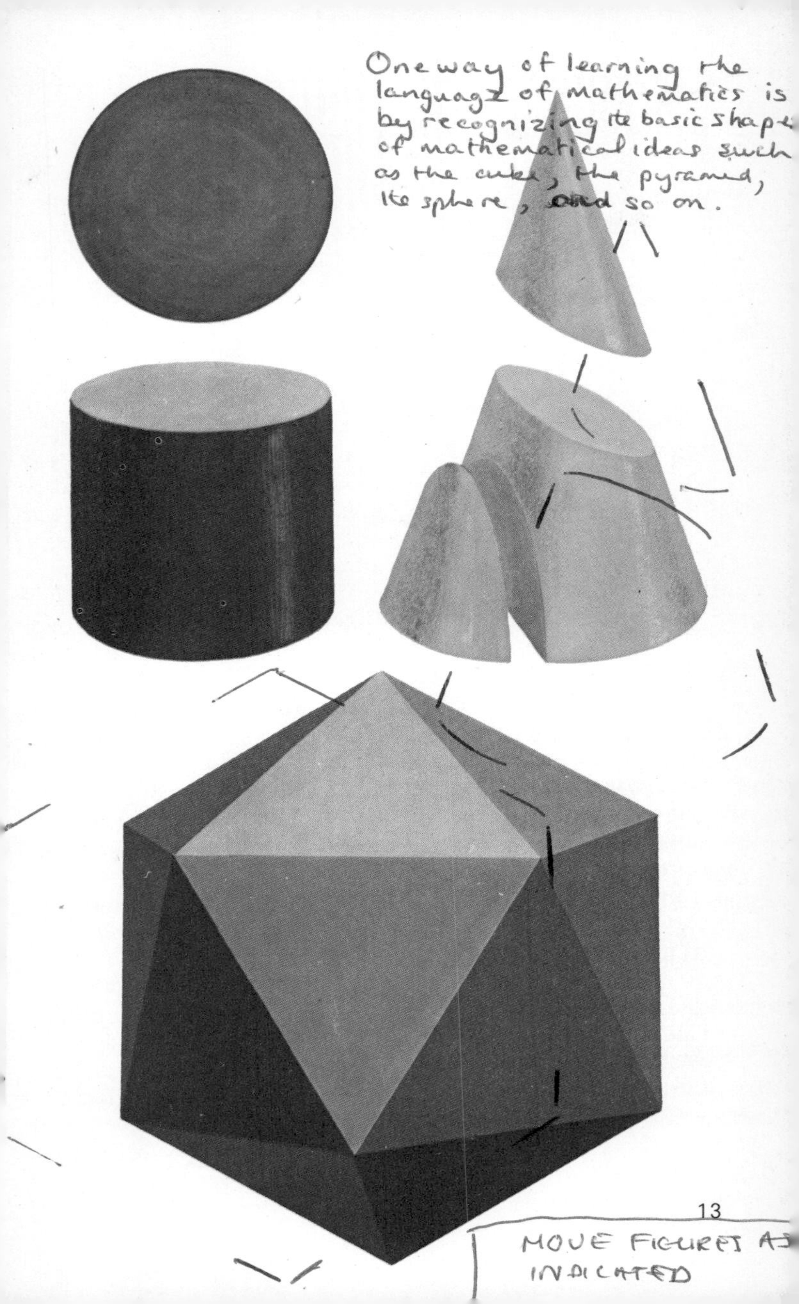
One way of learning the
languagz of mathematics is
by recognizing the basic shapes
of mathematical ideas such
as the cubz, the pyramid,
the sphere, and so on.
MOVE FIGURES AS
INDICATED

KINDS OF NUMBERS

'Bless us, divine number' – Plato

When we speak of 'numbers' we naturally tend to think first of the so-called 'natural' numbers, 1, 2, 3 and so on. But there are at least five kinds of numbers – some of them including other classifications, some being independent of their fellows.

In addition to the natural numbers we have the same symbols each with a minus sign prefixed. Together these make up the integers (positive and negative).

Then come the rational numbers. These comprise (together with the integers) all numbers that can be expressed as vulgar fractions or as finite or recurring decimals. Included among these are, for example, $\frac{2}{5}(\cdot 4)$, $\frac{1}{7}(\cdot\dot{1}4285\dot{7})$ – the latter being a recurring decimal; that is, a decimal in which a number or a series of numbers recurs in a set order.

Not included are numbers like $\sqrt{2}$ (the square root of 2), which when expressed decimally turns out to be 1·41421... The figures after the decimal point go on for ever, but without revealing any recognizable pattern. They are not 'rational' inasmuch as we cannot find a *precise* value for them, although we can come as near to exactitude as is necessary. If you wish to know the length of the diagonal of a square with a side of, say, one mile, your solution will be $\sqrt{2}$ miles; and if you take the approximation $\sqrt{2} = 1{\cdot}41421$ you vary from the right result by only about one part in five million – say a hundredth of an inch. It is difficult to think of any calculation in which this margin of error would be worth worrying about; however, if you did find yourself making a calculation which required this degree of accuracy, it would only be necessary to carry the sum on to a few more places of decimals.

All these numbers are, of course, 'real'. They can be plotted on a graph, as we shall see when we come to the section on 'Mathematics in Pictures'. That, however, is not true of the 'imaginary' numbers, such as $\sqrt{-1}$ (often written as i). You can, as we have seen, get as near as you like to accuracy in computing $\sqrt{2}$, but there is no way of finding even an approximate value for $\sqrt{-1}$ or of showing it on a graph. (Graphs are sometimes employed to *illustrate* the use of imaginary numbers; but that is not quite the same thing.)

This is a hurdle which many students find it difficult to surmount. Perhaps the simplest way is to regard $\sqrt{-1}$ and other 'imaginary' numbers as *operatives*.

When we multiply 4 by 3, 4 is the number *qua* number and 3 is the operative. In this particular case, of course, the distinction is unimportant, since 4×3 is the same thing as 3×4. In division that is not so; $10 \div 5$ is by no means the same thing as $5 \div 10$. The numbers 0 and

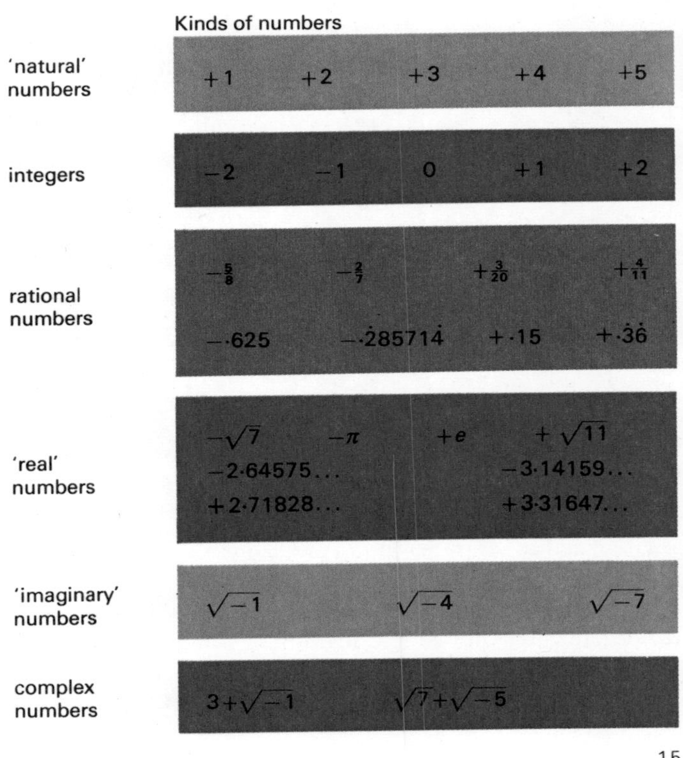

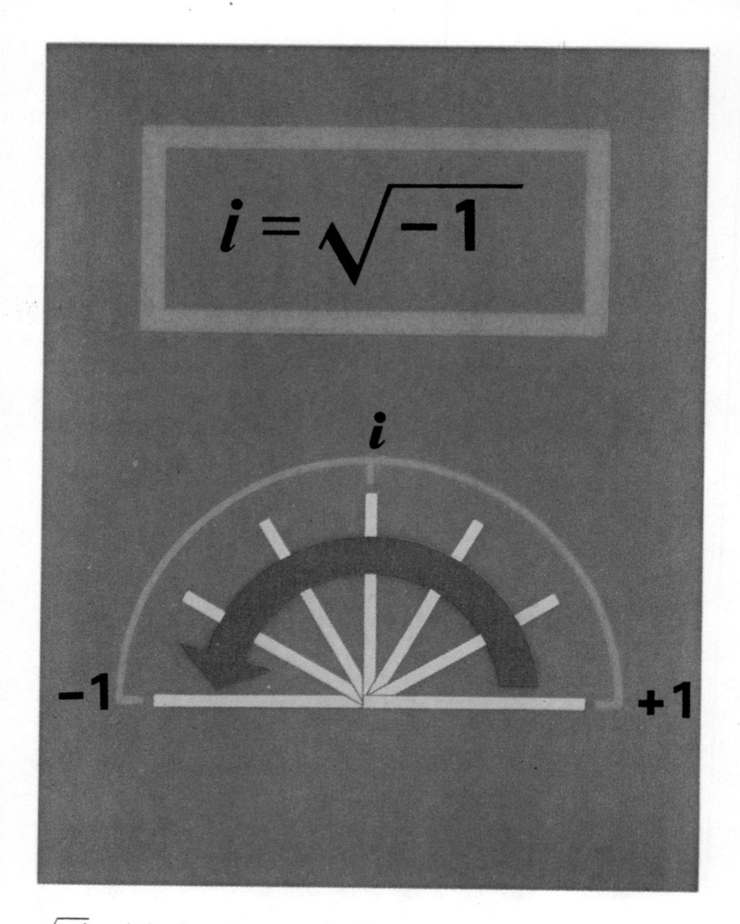

$\sqrt{-1}$ or i, the 'imaginary number', is represented here at the half-way point of an anti-clockwise progression between +1 and −1, when +1 has gone half-way to multiplying itself by −1.

infinity (∞) should never be used as operatives in basic arithmetic: $0 \times 3 = 0$ is a legitimate statement; 3×0 is essentially meaningless. Equally, $\infty \times 4 = \infty$ is a sound proposition; $4 \times \infty$ is a mathematical absurdity. What would be 0×0, or $\infty \times \infty$?

In dealing with wavelengths, electricians regard the movement of a straight line through half a circle in an anti-clockwise direction as travelling from a point on the right marked as '$+1$' to a point on the left marked '-1'. Clearly the operation can be represented as a process of multiplying by -1, which is the same thing as multiplying by $\sqrt{-1}$ *twice*. So that when the line has moved half-way – when it is perpendicular to its original position – it has gone half-way to multiplying itself by -1, and it is convenient to describe its new position with the expression $\sqrt{-1}$, or i.

It is true that $\sqrt{-1}$ does not exist in the sense that a loaf of bread exists. It is a man-made tool, and a very useful one. It has been contended by some mathematicians that God made the natural numbers, all the rest of the art being man's invention. That may or may not be so; but we *must* take the credit or bear the responsibility for $\sqrt{-1}$.

Complex numbers are the combination of 'imaginary' numbers with 'real' numbers, as, for instance, $6 + \sqrt{-5}$, or $\sqrt{2} + \sqrt{-1}$. They can be dealt with in just the same way as can other algebraic expressions. For example:

$$(4 + \sqrt{-3})(4 - \sqrt{-3}) = 19.$$

The explanation is as follows: we know

$$(x + y)(x - y) = x^2 - y^2,$$

so it follows that

$$\begin{aligned}(4 + \sqrt{-3})(4 - \sqrt{-3}) &= 4^2 - (\sqrt{-3})^2 \\ &= 16 - (-3) \\ &= 19.\end{aligned}$$

The reader may have noticed that every time I have referred to the 'imaginary' numbers I have put the adjective in inverted commas. The reason is that I think (and I am not alone in thinking) that it is a most unfortunate word to use, since *all* numbers are in a sense imaginary. You can see two men, eat two sandwiches, walk two miles and weigh two parcels; but who has ever seen, eaten, walked or weighed 'two'? Number, whether we describe it as 'real' or 'imaginary', is an abstraction; and the fact that it has many practical applications, and that it is indeed essential to all the physical sciences (and almost certainly to other sciences as well) should not blind us to that fact.

NUMERICAL SYSTEMS

'Observe how system into system runs' – Pope

We are so used to counting in sets of ten that we have almost come to regard the decimal system of notation as an immutable law of nature. Of course it is nothing of the kind, but merely the method we have adopted as being convenient (although it is by no means universally agreed to be the best possible). It is generally supposed that we count in tens because we have ten fingers (including the thumbs) and ten toes. Perhaps the three-toed sloth, if it learned to count at all, would do so in sets of six, the cloven-hoofed ox in sets of four, the horse would be put to the quite intolerable inconvenience of counting in sets of two and presumably the worm would do without counting altogether.

Having ten fingers and ten toes probably caused man to adopt decimal notation. If it could count at all, a three-toed sloth might use sets of six and a cloven-hoofed ox sets of four.

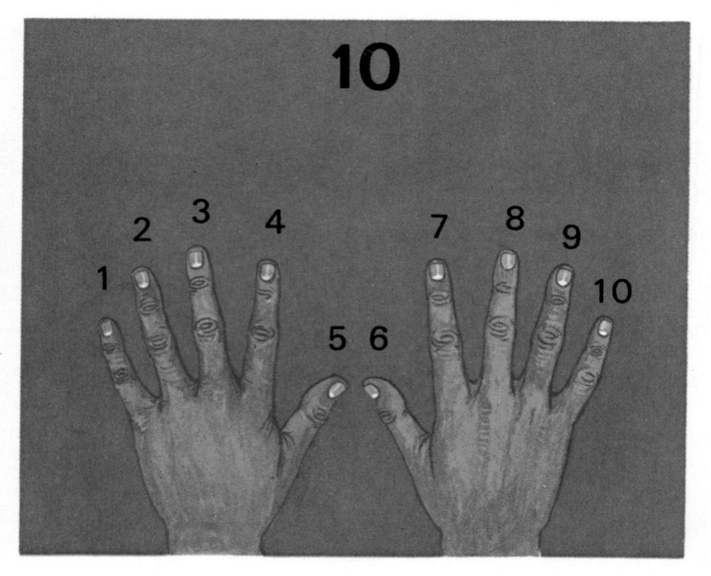

6
1 2 3 4 5 6 1 2 3 4 5 6
12 34 12 34
4

caustons – make this larger please.

Egyptian
Babylonian
Greek
Mayan

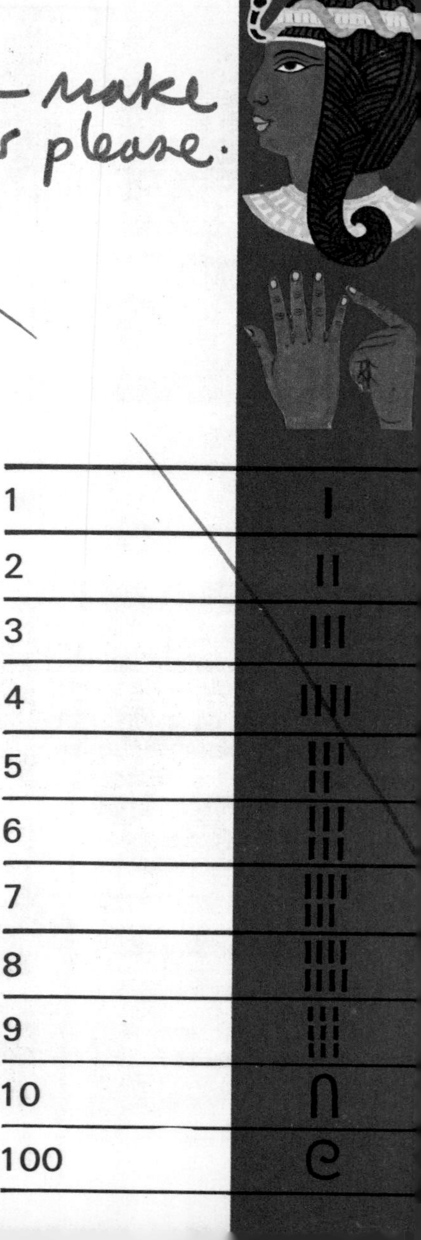

Early numerical systems were organized by grouping unit symbols together and by the invention of other symbols to represent larger groupings, of five, ten, sixty, a hundred and so on. Thus the Babylonian sign for 100, second column, comprises the symbol for sixty placed next to four 'ten' symbols ($60 + 40 = 100$). The Mayan symbol for 100, fourth column, is differently composed: here the 'five' symbol appears above that for twenty, denoting multiplication ($5 \times 20 = 100$).

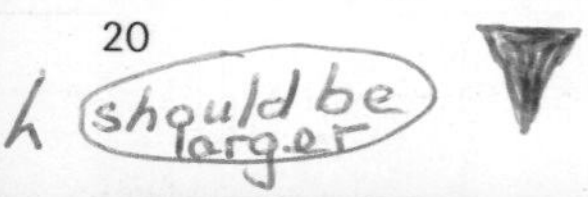

A

B

Γ

Δ

E

F

Z

H

Θ

I

P

It is worth noting that many people consider the duodecimal system (using sets of twelve) to be preferable to the decimal system (using sets of ten). It entails the use of two extra symbols, *t* and *e* to represent ten and eleven (10 and 11 in our notation) with 10 standing for our twelve. Once the strangeness has worn off, this is simple enough to operate. The advantage of the system is that the basic number (twelve) has more divisors than has ten. One can express in single-figure duodecimals halves, thirds, quarters, sixths and twelfths, whereas the decimal system allows only for halves, fifths and tenths. Compare the figures shown opposite. A half is expressible in either system with equal facility; of the other seven fractions five are simpler in duodecimals and only two in the decimal system.

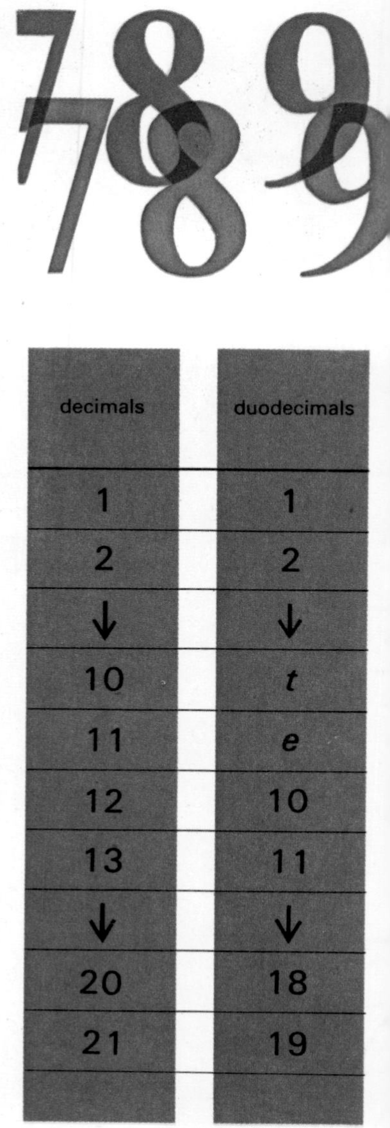

decimals	duodecimals
1	1
2	2
↓	↓
10	*t*
11	*e*
12	10
13	11
↓	↓
20	18
21	19

On this page the decimal system of numbers (counting in tens) is compared with the duodecimal system (counting in twelves) which entails the use of two extra symbols, *t* and *e*, to represent ten and eleven. On the opposite page vulgar fractions are given with their decimal and duodecimal equivalents.

vulgar fractions	decimals	duodecimals
$\frac{1}{2}$	·5	·6
$\frac{1}{3}$	$\cdot\dot{3}$	·4
$\frac{1}{4}$	·25	·3
$\frac{1}{5}$	·2	·2497
$\frac{1}{6}$	·16	·2
$\frac{1}{8}$	·125	·18
$\frac{1}{10}$	·1	·12497
$\frac{1}{12}$	·083	·1

Of course scales of notation can be constructed to any base. If, for example, one worked to the base of four, then 'natural' numbers corresponding to one to nine would be 1, 2, 3, 10, 11, 12, 13, 20, 21 ($21 = 2 \times 4 + 1 = 9$). Any of these systems will 'work'; the question is merely which of them is the most convenient. And on that basis twelve seems at least to merit consideration.

Another interesting scale of notation is the binary system, with the base of two, in which the only symbols used are 1 and 0. Thus one is, of course, 1; two is 10 (one set of two); three is 11 (one set of two and one unit); four is 100, and so on. This has the advantage of making ordinary calculations extremely simple and the disadvantage of making them cumbersome and time-consuming. The system is the basis of all electronic computers; first because, since they work almost instantaneously, the time factor is of practically no consequence, and secondly because electric current knows no middle course between 'on' and 'off', so that impulses corresponding with 1 and 0 (or with plus and minus, or 'yes' and 'no') are the only signals that electronic computers can convey.

The binary system is the secret of a little trick with which you may puzzle the less sophisticated of your friends. It is a method of multiplying two numbers together using no mathematical technique more difficult than adding, and multiplying and dividing by two. One writes down the two numbers side by side, then in consecutive lines multiplies the right-hand figure by two and divides the left-hand figure by two (ignoring all fractions – half of eleven, for example, would be written simply as '5' not as '5½'). Now cross through every line in which the *left-hand* number is even and sum what remains of the right-hand column. The total will give you the required result. In the illustration on page 26 the simplicity of this particular method of multiplication is demonstrated by solving the problem 41×13.

To show how the binary system operates in relation to the decimal system, the 'decimal' number 232 is here analysed and converted into its binary equivalent.

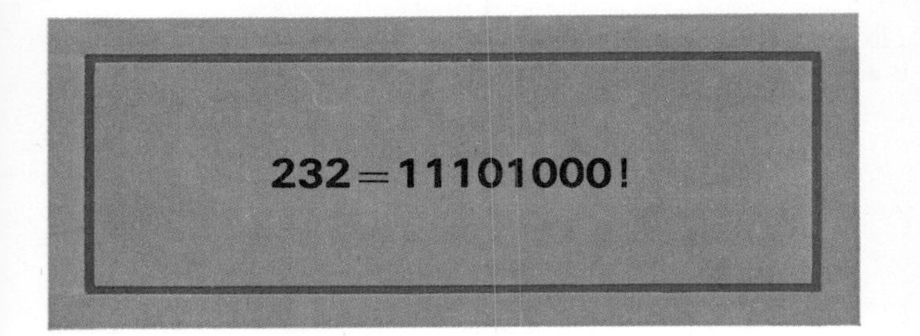
232=11101000!

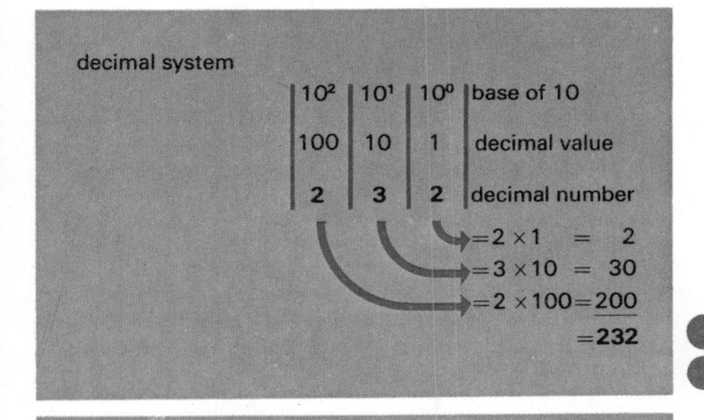
decimal system
10²
10¹
10⁰
base of 10
100
10
1
decimal value
2
3
2
decimal number
=2 × 1 = 2
=3 × 10 = 30
=2 × 100=200
=232

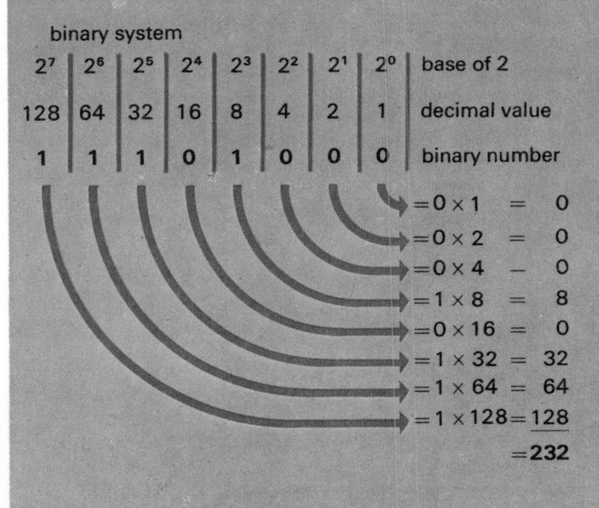
binary system
2⁷
2⁶
2⁵
2⁴
2³
2²
2¹
2⁰
base of 2
128
64
32
16
8
4
2
1
decimal value
1
1
1
0
1
0
0
0
binary number
=0 × 1 = 0
=0 × 2 = 0
=0 × 4 – 0
=1 × 8 = 8
=0 × 16 = 0
=1 × 32 = 32
=1 × 64 = 64
=1 × 128=128
=232

Multiplication by the binary system

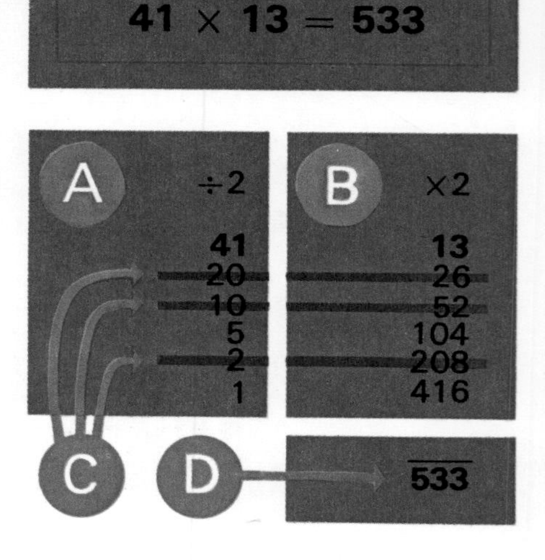

Only people familiar with the binary system will see how the trick works. Since any number can be written in the binary (or any other) system, it follows that any number must be the total of selected members of the series 1, 2, 4, 8, 16, 32 … (for example, 14 = 2 + 4 + 8, 21 = 1 + 4 + 16). To find the terms required for, say, 27, first note that it is odd, that is to say there will be a 1 in the unit column. We are left with 26, which is divisible by 2 but not by 4, so that there will be a 1 in the 'two' (2^1) column. That leaves 24, which is divisible by 4 and *also* by 8, so that there must be a 0 in the 'four' (2^2) column and a 1 in the 'eight' (2^3) column. The remainder is 16, giving a 1 in the 'sixteen' (2^4) column. Thus we have placed, reading from left to right, the figures 1, 1, 0, 1 and 1, so that 27 is in binary notation 11011 ($1 \times 16 + 1 \times 8 + 0 \times 4 + 1 \times 2 + 1 \times 1 = 27$).

What number heads the right-hand column is irrelevant. We *could* use the method to 'solve' the not too difficult problem of multiplying 41 by 1, see page 27, where our result (41) is made up of $1 + 8 + 32 = 1 + 2^3 + 2^5$.

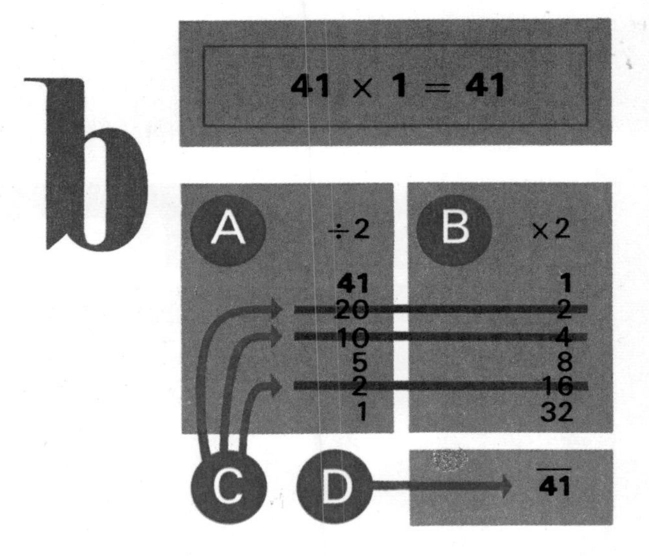

In the same way we can readily understand that 41 = $1 + (0 \times 2^1) + (0 \times 2^2) + (1 \times 2^3) + (0 \times 2^4) + (1 \times 2^5) =$ 101001 in the binary system; and the 'trick' that puzzled the uninitiated was merely crossing out the lines that would in the binary system require a 0 to be entered.

Many numerical tricks are based on the binary system. For example, to 'guess' the age of a friend, taking care to select one who has not yet reached his or her thirty-second birthday, make a card using the figures in the table overleaf.

Now ask your friend to tell you the column or columns in which his age appears, and you can at once tell him what that age is. You achieve the feat simply by adding together the top numbers of the columns indicated.

Why is the table compiled in this way? Remembering the binary system should enable you to answer the question easily enough. Why is the highest number in it 31? (We could equally well construct a shorter table going up to 15 or a longer one reaching as far as 63.) Each of these three numbers is one less than a power of 2. And with that hint you should be able to rationalize the list.

Head the five columns with the numbers 1, 2, 4, 8 and 16. Let the first column contain all odd numbers up to 31, the second alternate pairs of numbers (2 and 3, 6 and 7, etc.), the third alternate quartets (4, 5, 6 and 7, 12, 13, 14 and 15, etc.), the fourth octets (8 to 15 inclusive and 24 to 31 inclusive), and the fifth the numbers from 16 to 31.

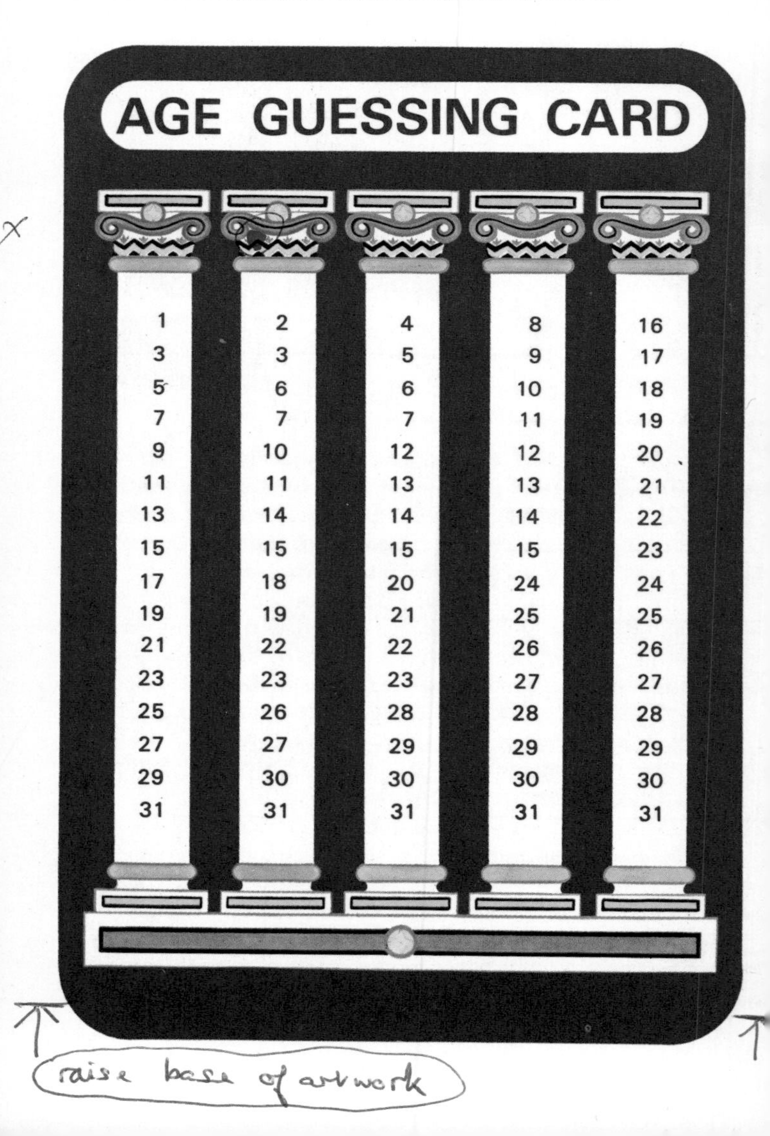

Once more the explanation of the trick is that since any number can be expressed in the binary system it must be the sum of selected numbers from the series 1, 2, 4, 8, etc. (2^0, 2^1, 2^2, 2^3, etc.).

Taking a number at random, say, 22, this is equal to 2 + 4 + 16, and in binary notation is written 10110. Reading these figures in reverse order shows that the required number will not appear in the 2^0 column (headed by unity), will be in the second and third columns (headed respectively by 2 and 4), will not be in the fourth, but appears again in the fifth (headed 16). The number 31, being the sum of 1, 2, 4, 8 and 16, of course appears in all the columns.

The trinary system is another notation worth mentioning (its numerals being 1, 2, 10, 11, 20, and so on) because its use provides the key to an amusing little problem.

It has already been pointed out that any number can be built up by adding selected members of the binary system. In the trinary system we cannot do this by selecting from 1, 10, 100, and so on (corresponding to our 1, 3, 9, etc.). But it *can* be done if we also permit ourselves the use of minus numbers, for example ... − 100, − 10, − 1, + 1, + 10, + 100 ... (corresponding to our ... −9, −3, −1, + 1, + 3, + 9 ...).

This leads us back to the problem I mentioned before. If you wanted to find four pound-weights which would enable you to weigh any exact number of pounds from one to forty, which four weights would you choose? Of course you would need weights with values of 1, 3, 9 and 27 pounds (the decimal equivalents of the trinary 1, 10, 100 and 1,000) because by using either the 'plus' or 'minus' values of these four numbers you could determine the weight of any object weighing less than our limit of forty pounds. All you would need to do would be to put the 'minus' weights in one pan of the scale together with the object to be weighed, and the 'plus' weights in the other. Let us suppose, for example, that the given object weighs 32 pounds. Now 32 = 27 + 9 − 3 − 1 so that one pan of the scale will contain the 32 pound object, plus the 3 and 1 pound-weights (making a total of 36 pounds), and the other will contain the 27 and 9 pound-weights, which also make a total of 36 pounds.

QUADRATIC EQUATIONS

'Mind your P's and Q's' – Old proverb

Quadratic equations are those which contain the term x to the power of two but of no higher power. They should occasion no difficulty, since there is a convenient formula by which they can be solved, whether their roots are 'real' or complex. The only trouble is remembering the formula.

First, any quadratic can always be expressed in the form $ax^2 + bx + c = 0$. No matter how complicated it may be in its original form (take, for example:

$$-78\tfrac{1}{2}x^2 + 222x = 17\tfrac{2}{3})$$

the transposition is easy enough; you merely take $a = -78\frac{1}{2}$, $b = +222$, $c = -17\frac{2}{3}$ and proceed accordingly. The formula is:

$$x = \frac{-b \pm \sqrt{b^2 - 4ac^*}}{2a}.$$

He is a poor mathematician who is satisfied to use a formula merely because it 'works'. He should wish to know *why* it produces the required result.

Any quadratic expression has two factors, 'real' or complex. Sometimes you can recognize them at a glance: it is not hard to see that $x^2 - 5x + 6 = (x - 2)(x - 3)$, and hence that if $x^2 - 5x + 6 = 0$, x must be equal to either $+2$ or $+3$ (because if two numbers multiplied together make 0, then at least one of them must *be* 0). Unfortunately, things don't always work out so simply.

Given that $ax^2 + bx + c = 0$, we can divide both sides of the equation by a without invalidating it, so that $\frac{ax^2}{a} + \frac{bx}{a} + \frac{c}{a} = 0$. Now, as any quadratic expression has two factors, we may say $x^2 + \frac{bx}{a} + \frac{c}{a} = (x - p)(x - q)$, although we do not as yet know the values of p and q. And $(x - p)(x - q) = x^2 - x(p + q) + pq = 0$. So that $x^2 - x(p + q) + pq = x^2 + \frac{bx}{a} + \frac{c}{a}$; and we shall have

* See also Appendix 1: 'Extraction of Roots', page 150.

INDENT 1 EM

achieved our factorization and thus solved the equation if we put $(p + q) = -\frac{b}{a}$ and $pq = +\frac{c}{a}$.

From this it follows that:

$$(p + q)^2 = p^2 + 2pq + q^2 = \frac{b^2}{a^2}$$

and

$$4pq = \frac{4c}{a}.$$

So, subtracting, $p^2 - 2pq + q^2 = \frac{b^2 - 4ac}{a^2}$

and $\sqrt{p^2 - 2pq + q^2} = p - q = \pm \frac{\sqrt{b^2 - 4ac}}{a}$.

But we already know that $p + q = -\frac{b}{a}$.

So,

$$p + q = -\frac{b}{a}$$

$$p - q = \pm \frac{\sqrt{b^2 - 4ac}}{a}$$

and, again subtracting,

$$2q = \frac{-b \pm \sqrt{b^2 - 4ac}}{a}$$

so that

$$q = \frac{-b \pm \sqrt{b^2 - 4ac}}{2a}$$

and

$$p = \frac{-b \mp \sqrt{b^2 - 4ac}}{2a}.$$

But since $(x - p)(x - q) = 0$, one of these two expressions must equal 0, so that $x = p$ or $q = \frac{-b \pm \sqrt{b^2 - 4ac}}{2a}$.

Try it out with the very easy quadratic we instanced before: $x^2 - 5x + 6 = 0$. This gives $a = +1$, $b = -5$, $c = +6$.

So $x = \frac{+5 \pm \sqrt{25 - 4.6}}{2} = \frac{+5 \pm 1}{2} = +3$ or $+2$.

$$3^2 - 5.3 + 6 = 9 - 15 + 6 = 0.$$
$$2^2 - 5.2 + 6 = 4 - 10 + 6 = 0.$$

There are formulae, though very much more complicated ones, for solving cubic equations (which have three roots) and quartics with four. For equations of even higher order, unless you are lucky enough to find a factor, you must use graphs or an electronic computer.

INDICES AND LOGARITHMS

'Logarithms are numbers invented for the more easy working of questions in arithmetic and geometry' – Henry Briggs

Most people are familiar with indices – the little symbols above the line that show the power to which the main figure is to be raised. They know, for example, that $2^3 = 2 \times 2 \times 2 = 8$; $3^5 = 3 \times 3 \times 3 \times 3 \times 3 = 243$; $5^2 = 5 \times 5 = 25$. It is, however, essential to know rather more than that about indices before one can go on to logarithms, one of the most outstanding 'short cuts' ever devised in mathematics.

It is clear that if you wish to multiply, say, 2^5 by 2^7, 2^5 can be written as 2.2.2.2.2 and 2^7 as 2.2.2.2.2.2.2. So their product will obviously be 2.2.2.2.2.2.2.2.2.2.2.2, or $2^5.2^7 = 2^{12}$. Here we see that the *multiplication* of these two numbers has been achieved by merely *adding* their indices; in other words, $2^5.2^7 = 2^{5+7} = 2^{12}$.

Clearly this applies not merely to twos but to any numbers, as can readily be seen:

$$x^3 = x.x.x$$
$$x^5 = x.x.x.x.x$$
$$x^3.x^5 = x^8 = x.x.x.x.x.x.x.x = x^{3+5}$$

or, more generally, $x^a.x^b = x^{a+b}$. In the same way, of course, $x^a \div x^b = x^{a-b}$.

The value of this is that it enables us to do quite a number of sums in multiplication by means of addition (a much simpler process). When we look at logarithms we shall see that we can (if it is worth the trouble) do *any* multiplication sum in this way.

If a number is to be squared you multiply its index by two, if to be cubed you multiply the index by three. To find the square root of a number you divide its index by two; for the cube root, you divide by three. For example: $\sqrt{3^4} = 3^{4/2} = 3^2$; $\sqrt[3]{x^6} = x^{6/3} = 6^2$. In general,

$$x^a \times x^b = x^{a+b}$$
$$x^a \div x^b = x^{a-b}$$
$$(x^a)^b = x^{ab}$$
$$\sqrt[b]{x^a} = x^{a/b}$$

So long as we are dealing with integers this is not very difficult; one can readily see that $x^5 \div x^3 = x^{5-3} = x^2$, and that $\sqrt{x^6} = x^3$. But what happens when the result of this process turns out to be fractional or negative?

The answer is: nothing very sensational. We proceed in exactly the same way: $x^5 \div x^7 = x^{-2}$; $\sqrt{x^3} = x^{1\frac{1}{2}}$.

Expressions of this kind are often a stumbling block for students. We know very well what is meant by x^2: it is 'multiply two x's together'. But how do you multiply *minus* two x's together? And, forgetting for a moment about the negative signs, how do you multiply one-and-a-half x's together? What indeed *are* one-and-a-half x's?

An early calculating device for multiplying numbers, known as Napier's rods, was invented in 1617 by John Napier, a Scotsman.

The answer here is comparable with that given in regard to imaginary numbers. Numbers are man-made tools; they justify their existence if they answer the pragmatic test – if they *work*. A hammer is a useful tool for putting nails into wood; we need not philosophize to explain *why* we usually make hammers with wooden handles and iron heads. Sufficient to say that we find it convenient.

Indices are in fact no more mysterious than hammers. When we say that $x^2 \div x^2 = x^{2-2} = x^0$, that is an entirely proper and meaningful statement, although nobody, I imagine, could tell you precisely what is meant by multiplying something by itself no times. Nevertheless, it works: $6^3 \div 6^3$ is clearly one, and $6^3 \div 6^3 = 6^{3-3} = 6^0$. In fact $n^0 = 1$ whatever n may be.

Another machine, Pascal's calculator, was invented in 1642 by the French philosopher and mathematician, Blaise Pascal. This device enabled the user to add figures to six places, automatically carrying the tens.

In the same way, let us not be too terrified of fractional indices. We have already said that $\sqrt{x^n} = x^{n/2}$. Now let us try this *en clair*. $\sqrt{6^4} = \sqrt{1296} = 36 = 6^2 = 6^{4/2}$; $\sqrt[3]{2^6} = \sqrt[3]{64} = 4 = 2^2 = 2^{6/3}$.

So fractional indices are clearly devices we can *use*. Let us try to see a little more clearly what they mean. What is, for example, $x^{2\frac{1}{2}}$?

We have seen that $\sqrt{x^n} = x^{n/2}$. Now $x^{2\frac{1}{2}} = x^{5/2}$, and this must clearly be $\sqrt{x^5}$. Here we find that fractional indices have a perfectly plain and 'real' value. To find $x^{2\frac{1}{2}}$, multiply five x's together and take the square root of the result. $9^{2\frac{1}{2}} = \sqrt{9.9.9.9.9} = \sqrt{59049} = 243 = 9.9.3 = 9.9.\sqrt{9} = 9.9.9^{1/2} = 9^{2\frac{1}{2}}$.

With negatives there are still no complications. The value of x^{-3} is clearly $x \div x^4 = x^{1-4} = x^{-3}$. But $\frac{x}{x^4} = \frac{1}{x^3}$; so that we arrive at the conclusion that $x^{-n} = \frac{1}{x^n}$, and this will be found always to give the right answer.

Now, a logarithm is simply the power to which a given number (the base) must be raised to produce a required result. For example, if we take 2 as the base, $2^3 = 8$, so that log 8 to the base 2 is 3. Note that because $2^0 = 1$, log 1 to the base 2 (or indeed to any other base) must be 0.

How does this help? Simply enough. Suppose we are asked to multiply 8 by 8 (admittedly not a very arduous task). We can do it (and of course should do it) by merely remembering our eight times table. But we can if we like equally well say $8 = 2^3$, so $8 \times 8 = 2^{3+3} = 2^6 = 64$. Alternatively, log 8 to the base 2 = 3, so log 8 × log 8 = 3 + 3 = 6, and therefore $8 \times 8 = 2^3 \times 2^3 = 2^6 = 64$.

All this seems rather like trying to use a sledgehammer to crack a nut, because we are applying logarithms (and not the best kind, because 2 is a very unsatisfactory base) to sums which stand in no need of such treatment. You may recall that in *In Good King Charles's Golden Days* Shaw makes Isaac Newton use logarithms to work out the price of a dozen herrings. Obviously Newton would never have done anything of the kind; but the solution at which he arrived agreed with the sum demanded by the fishmonger's boy, and thus may be taken as being correct.

Logarithms are normally calculated to the base 10.* That is to say, the logarithm (normally contracted to log) of a number n is the power to which 10 would have to be raised to make the result n. In other words, if $x = \log n$, then $10^n = x$. It is at once clear that if $n = 1$, $x = 10$ (so that $\log 10 = 1$), and if $n = 0$, $x = 1$ ($\log 1 = 0$). But of course this doesn't carry us very far.

Supposing we wish to construct a table of logarithms to the base 10 (fortunately we don't have to do so, since such tables are readily available; but we shall understand logarithms better if we make a start at doing so). We know to begin with, of course, that $\log 1 = 0$, $\log 10 = 1$, $\log 100 = 2$, and so on. But clearly we must find other logarithms to fill up these very extensive gaps.

Now, $2^{10} = 1{,}024$ and $10^3 = 1{,}000$. These two numbers are not very distant from each other, so for a first approximation (just as we did in square root and other mathematical exercises) we may take it that 10^3 is roughly equivalent to 2^{10}. Thus if we take 2^{10} as being about the same thing as 10^3, we can go on to say that $10 \log 2$ will be not very far distant from 3, so that an approximate value for $\log 2$ will be $\cdot 3$. If you look up the logarithm of 2 in a table of four-figure logs you will find it given as $\cdot 3010$, which is not very far from our figure.

Again, $3^9 = 19{,}683$, which is not so very different from 20,000. So $\log 20{,}000$ is roughly equal to $\log 2 + \log 10{,}000$. We have already shown $\log 2$ to be roughly equal to $\cdot 3$, and of course $\log 10{,}000$ is 4. So that $\log 20{,}000$ (which is fairly close to 3^9) is about $4\cdot 3$. Then, since $\log 3^9 = 4\cdot 3$, it follows that $9 \log 3 = 4\cdot 3$, so that $\log 3 = \cdot 48$ approximately (a four-figure log table gives it as $\cdot 4771$, so again we are not far out).

Furthermore, since we know that $\log xy = \log x + \log y$, it is clear that $\log 2 + \log 5 = \log 10$, which we know to be one. And since $\log 2$ is approximately $\cdot 3$, $\log 5$ will be about $1 - \cdot 3 = \cdot 7$. Again this is quite close, $\log 5$ being shown as $\cdot 6990$ in the four-figure table.

* Napierian, or 'natural' logarithms, are calculated to the base e ($2\cdot 71828 \ldots$). We need not here be concerned with them.

We can now begin to construct a rough table on the basis of what we have already discovered, see the table below.

Thus, to multiply 3·16 by 31·6, we add the logarithms and take the antilogarithm of their total. Log 3·16 = ·5, log 31·6 = 1·5. Their sum is 2, of which the antilogarithm is 100. 3·16 multiplied by 31·6 in fact comes to 99·856. With an error of less than ·002 this is not too bad for what is admitted to be only a first approximation.

Be careful to remember that

$$\log x + \log y = \log xy$$
$$\log x - \log y = \log \frac{x}{y}.$$

As I have said, there is fortunately no need for you to construct your own log tables; you will find them in most mathematics books. In using them, however, you must bear in mind that for some reason (obscure, I must admit, to me) you must before looking up your number multiply it by ten: that is to say, if you want to find log 2 you look in your table for log 20; and if you want, for example, the logarithm of 3·4 then you look for log 34.

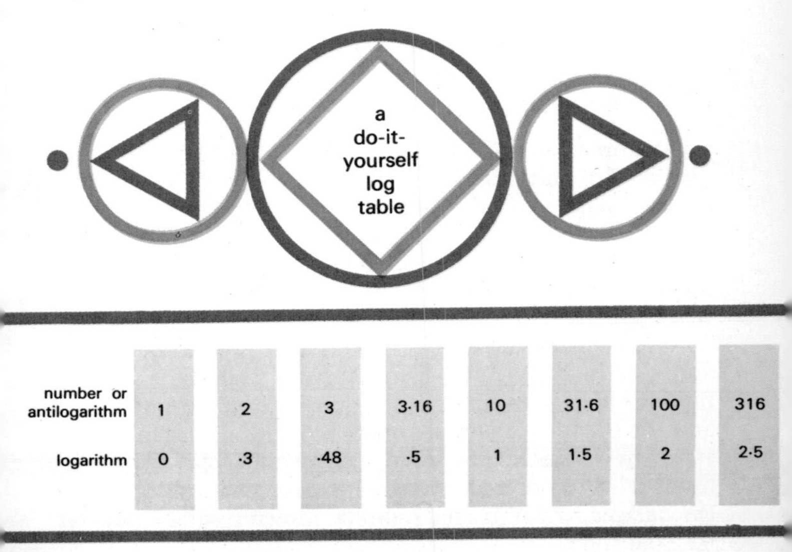

number or antilogarithm	1	2	3	3·16	10	31·6	100	316
logarithm	0	·3	·48	·5	1	1·5	2	2·5

LOGARITHMS

	0	1	2	3	4	5	6	7	8	9	1	2	3	4	5	6	7	8	9
10	·0000	0043	0086	0128	0170	0212	0253	0294	0334	0374	4	8	12	17	21	25	29	33	37
11	·0414	0453	0492	0531	0569	0607	0645	0682	0719	0755	4	8	11	15	19	23	26	30	34
12	·0792	0828	0864	0899	0934	0969	1004	1038	1072	1106	3	7	10	14	17	21	24	28	31
13	·1139	1173	1206	1239	1271	1303	1335	1367	1399	1430	3	6	10	13	16	19	23	26	29
14	·1461	1492	1523	1553	1584	1614	1644	1673	1703	1732	3	6	9	12	15	18	21	24	27
15	·1761	1790	1818	1847	1875	1903	1931	1959	1987	2014	3	6	8	11	14	17	20	22	25
16	·2041	2068	2095	2122	2148	2175	2201	2227	2253	2279	3	5	8	11	13	16	18	21	24
17	·2304	2330	2355	2380	2405	2430	2455	2480	2504	2529	2	5	7	10	12	15	17	20	22
18	·2553	2577	2601	2625	2648	2672	2695	2718	2742	2765	2	5	7	9	12	14	16	19	21
19	·2788	2810	2833	2856	2878	2900	2923	2945	2967	2989	2	4	7	9	11	13	16	18	20
20	·3010	3032	3054	3075	3096	3118	3139	3160	3181	3201	2	4	6	8	11	13	15	17	19
21	·3222	3243	3263	3284	3304	3324	3345	3365	3385	3404	2	4	6	8	10	12	14	16	18
22	·3424	3444	3464	3483	3502	3522	3541	3560	3579	3598	2	4	6	8	10	12	14	15	17
23	·3617	3636	3655	3674	3692	3711	3729	3747	3766	3784	2	4	6	7	9	11	13	15	17
24	·3802	3820	3838	3856	3874	3892	3909	3927	3945	3962	2	4	5	7	9	11	12	14	16

For log 2·314, remembering to multiply by ten, find 23 in the first column and take the number in the second column across, headed 1. To this number, ·3636, add the figure 7 appearing under column 4 in the table on the extreme right. So log 2·314 = ·3643.

A sample page of a log table is shown in the illustration above. To find log 2·314 you look up 23 in the first column, then, since the number is 2·31 . . ., take the second column on the right (headed 1), and finally (to account for the 4 in the third decimal place) add the figure 7 appearing under the heading 4 in the right-hand table, thus giving log 2·314 = ·3643.

A logarithm consists of two parts, the characteristic and the mantissa, the characteristic being the integral number that appears on the left of the decimal point and the mantissa that part of the logarithm that follows it. It is the characteristic of the logarithm that shows the number of significant (integral) figures that will appear in the answer: it will be the number shown by the characteristic plus one.

Let us take an example: to multiply 4,573 by 6,089. Consulting a table of four-figure logs (there are more extensive ones if you want them), we find that

$$\log 4{\cdot}573 = {\cdot}6602$$
$$\log 6{\cdot}089 = {\cdot}7845.$$

So

$$\log 4{,}573 = 3{\cdot}6602$$
$$\log 6{,}089 = 3{\cdot}7845.$$

Therefore $\log 4{\cdot}573 + \log 6{\cdot}8089 = 3{\cdot}6602 + 3{\cdot}7845$
$= 7{\cdot}4447.$

$$\text{Antilog}\ {\cdot}4447 = 2{\cdot}78.$$

The fact that the characteristic of the logarithm is 7 shows that there will be 8 significant figures in the answer. So the product should be about 27,800,000.

Working the sum out the 'hard way', we find that $4{,}573 \times 6{,}089 = 27{,}844{,}997$. A fair approximation; it would be closer, of course, if we had used a table of six-figure logs. And it has, after all, saved a little time.

You are not likely to come across a great many calculations in which you will need logarithms; but they can be useful. You may remember from your schooldays the old problem, if a man receives a penny for the first square on a chessboard, twopence for the second, fourpence for the third, and so on doubling as far as the 64th square, how much will he receive? This is excessively tedious to work out by a series of multiplications, but an approximation is easily arrived at by logarithms.

His total will be (in pence) $2^{64} - 1$. Forgetting for the moment the one penny that has to be deducted, we may say that he receives 2^{64} pence.

If $x = 2^{64}$,

$$\log x = \log 2^{64} = 64 \log 2 = 64\ ({\cdot}3010)$$
$$= 19{\cdot}2640.$$

$$\text{Antilog}\ {\cdot}2640 = 1{\cdot}8436,$$

and since the characteristic is 19 there will be 20 significant figures in the answer.

Therefore the man will receive something in the region of 18,400,000,000,000,000,000 pence, or, if we convert to pounds, about £76,700,000,000,000,000. For the precise, his actual receipts should be 18,446,744,073,709,551,615 pence, or £76,861,433,640,456,465 1s. 3d.

The slide-rule, which is illustrated below, is really nothing more than a mechanized table of logarithms. It is useful as a rough-and-ready method of calculation; but its value is limited by the impossibility of marking and reading very small divisions on a scale.

You can make your own rough slide-rule by taking two strips of cardboard and marking each of them with the numbers 1, 2, 3, 5, 7 and so on (for reasons that should become apparent, it is only necessary to mark prime numbers), making the space between 1 and 2 some multiple of log 2, that between 1 and 3 the same multiple of log 3, and so on. Then if, as shown in the illustration, you shift the lower scale so that number 2 comes under number 1, any number on the upper scale will correspond with twice that number on the lower scale. Again, if you shift the lower scale so that 3 comes under 1, any number on the upper scale will correspond with three times that number on the lower scale, and so on.

The logarithms of 2, 3, 5 and 7 respectively are ·3010, ·4771, ·6980 and ·8451. Suppose we multiply each of these

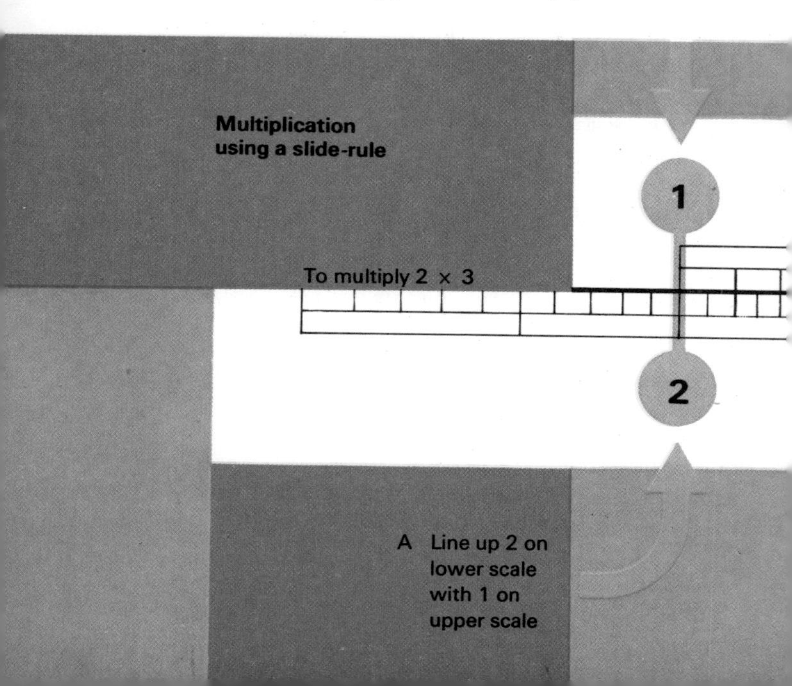

by 10, then the distance between 1 and 2 will be roughly 3 inches; that between 1 and 3, 4·8 inches; 1 and 5, 7 inches; and 1 and 7, 8·5 inches. Now, if we shift the lower scale so that 2 comes under 1, we know that 4 must come under 2, enabling us to fix the point 4 on both scales, and hence 6 and 8. In the same way, shifting the scale until 3 comes under 1 enables us to mark in all the necessary multiples of 3, and so on.

Equally we can find $1\frac{1}{2}$ on the upper scale by marking (when multiplying by 2) the point on the upper scale corresponding with 3 on the lower scale; and this process can continue as long as we please.

There are sophisticated variations of the slide-rule which allow more accurate and more complicated mathematical processes. One, allowing for readings correct to four or even five places of decimals, has an extra-long scale wound in a coil round a cylinder.

All these devices are of great practical value although the 'pure' mathematician may prefer (as I do) to rely on methods which though slower are more accurate.

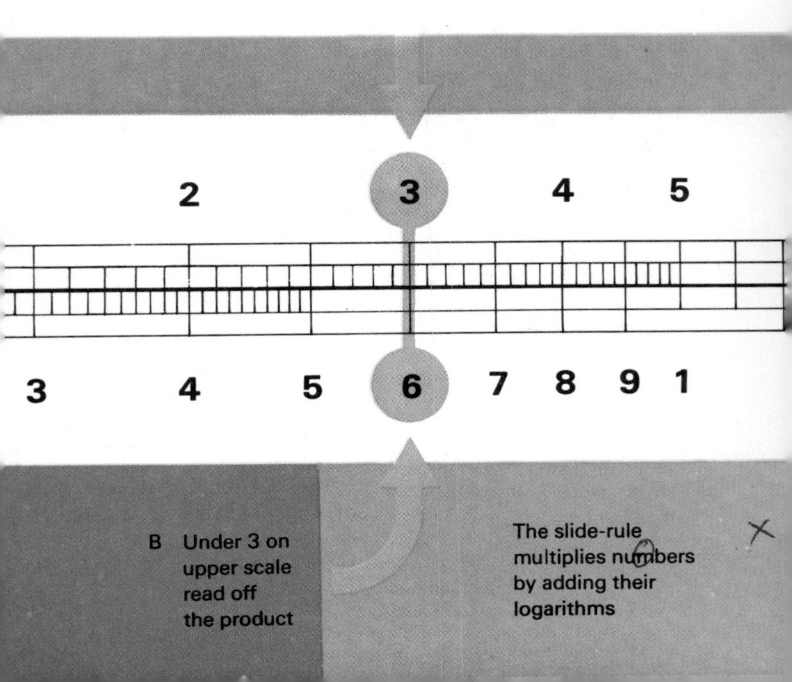

The slide-rule multiplies numbers by adding their logarithms

THE ART OF MEASURING

'Geometry, . . . the only science that it hath pleased God hitherto to bestow on mankind' – *Thomas Hobbes*
'God ever geometrizes' – *Plato*

It is sad to know that the derivation of the word 'arithmetic' from the Latin *ars metrica* (the art of measuring) is not etymologically correct. But if it be bad Latin it is nevertheless good mathematics. Mathematics is almost wholly concerned with measuring – distances, areas, volumes, weights, forces, whatever physical phenomena you will; and mathematicians do not as a rule wander out of their chosen field into the misty wastes of metaphysics. So-called mathematical 'proofs' of the existence or non-existence of God are not worth the paper they are written on. Indeed, it may be argued that a god whose existence was mathematically demonstrable would be a wholly unsuitable object of worship.

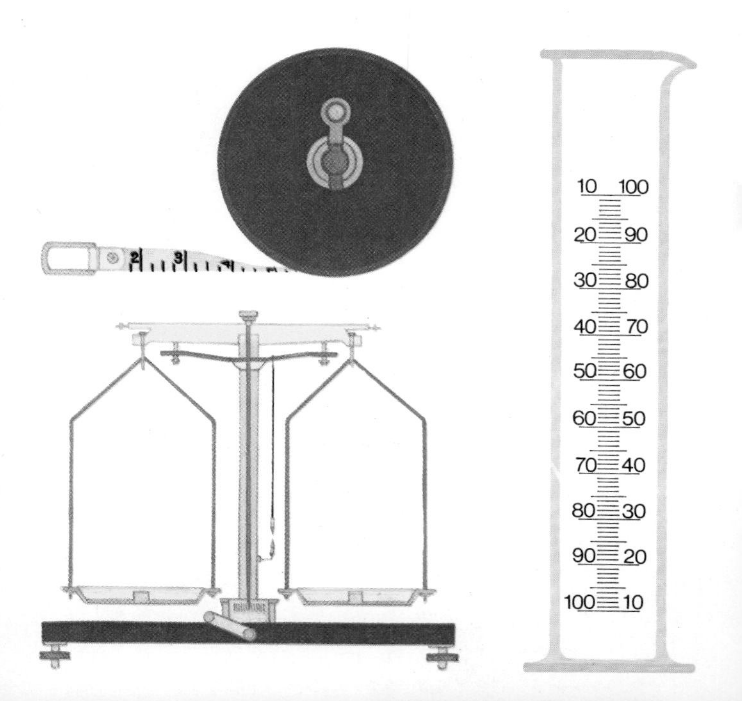

0
5
10
15
20
25
30
C

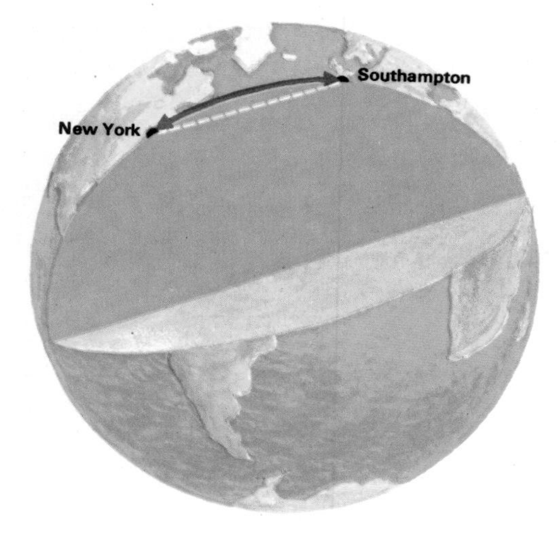

The shortest distance on the surface between Southampton and New York is not a straight line, but an arc of a great circle, the centre of which is at the centre of the earth.

A very large proportion of measuring is based on Euclid's plane geometry, which most of us learned at school and which is still taught at most schools today. It is now the fashion to sneer at Euclid and to say that his geometry has been superseded by more widely embracing systems.

To some extent this is true. Several of Euclid's 'axioms' (modern geometers refer more cautiously to 'postulates') have been found to be *not* self-evident, so that not all of his proofs are impeccable. But they do for the most part answer the pragmatic test – they 'work', or at least they work with sufficient accuracy to be a useful and indeed indispensable tool.

Also we must remember that Euclid's plane geometry is, as its name clearly shows, concerned with planes and not with curved surfaces. We live on a planet which is roughly spherical in shape, so that there are no plane surfaces within our experience, and the results given by

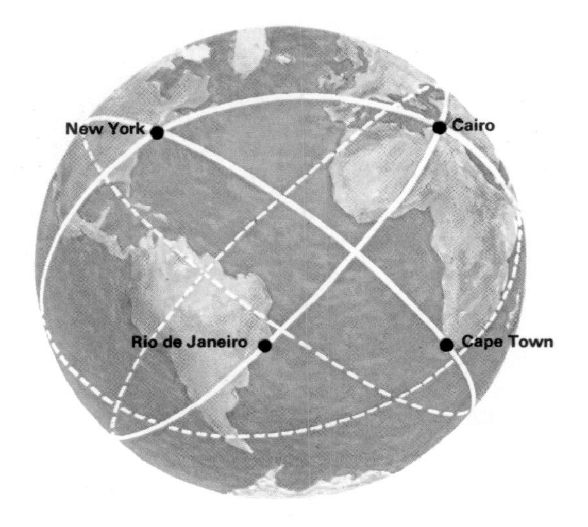

An unlimited number of great circles may be drawn round the earth and any two points that you can mention on the earth's surface are bound to be joined by one of them.

plane geometry need adjustment before they can be applied to curved surfaces. Ten yards measured on the earth's circumference do not extend quite as far as would ten yards measured 'on the flat'; but the difference is too small to be worth worrying about. It makes a considerable difference, however, when we are considering, for example, a long sea journey. We remember from our Euclid that the shortest distance between two points is a straight line. But the shortest distance between Southampton and New York is certainly not a straight line (unless we propose to get there by burrowing under earth and ocean). It is an arc of a great circle (that is, a circle having its centre at the centre of the earth). There is an unlimited number of great circles, and any two points on the earth's surface are both necessarily situated on one of them.

Another difficulty with Euclid's geometry is that he was concerned with ideals rather than with reality. A point, according to Euclid, has no dimensions; and as far as we

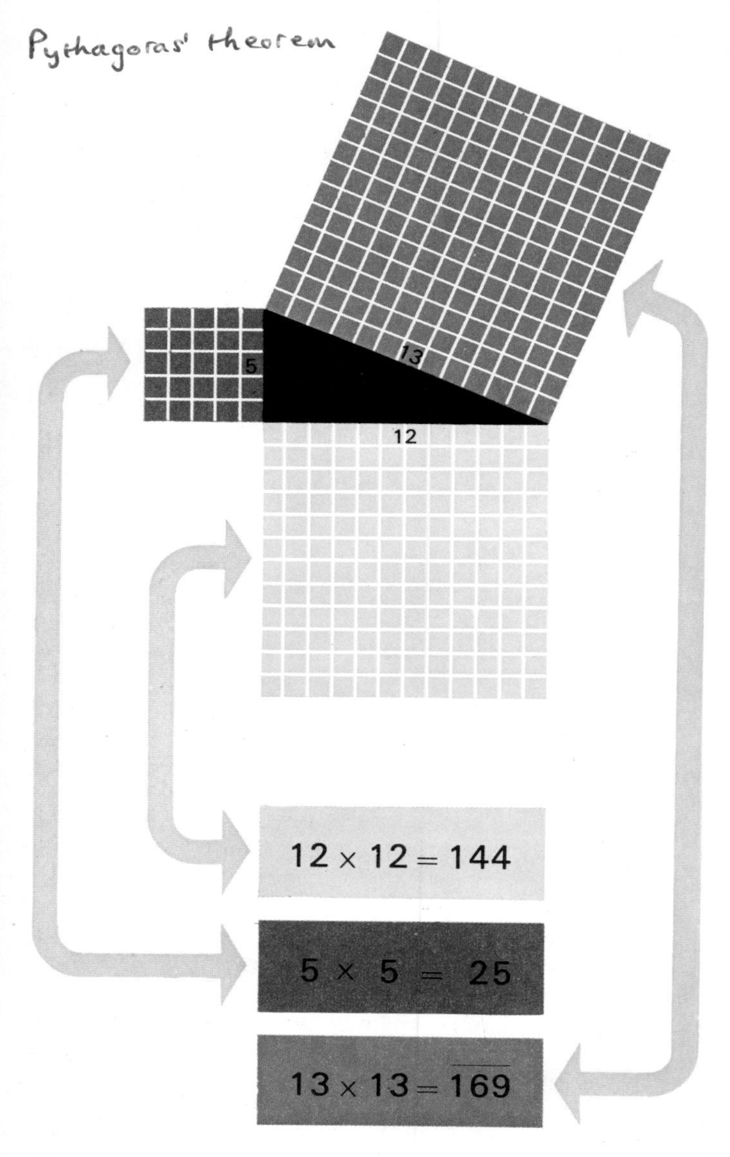
Pythagoras' theorem
5
13
12
12 × 12 = 144
5 × 5 = 25
13 × 13 = 169

are concerned, although we know of many tiny things, an object with *no* dimensions simply doesn't exist. Again, according to Euclid, a line has length but no breadth or thickness – and hence is by definition invisible. So, quite obviously, when the housewife hangs her washing on a 'line' she is not doing at all the same thing as carrying out the Euclidean instruction to 'draw a straight line from the point A to the point B'.

Be wary, by the way, of saying, perhaps somewhat censoriously, that the world doesn't 'live up' to Euclid's 'ideal' system. The word 'ideal' is used in the sense not of 'perfect' but of 'notional'; it depends on ideas rather than on facts. Where the facts do not correspond with an 'ideal' system it is the system, not the world, that must be regarded as imperfect.

With all its failings, however, an enormous amount of mathematical work is based on Euclid's geometry and in particular on one theorem, attributed to Pythagoras, that in any right-angled triangle the square on the hypotenuse is equal to the sum of the squares on the other two sides; see the illustration opposite which demonstrates the notion pictorially.

For those who have forgotten the proof of the theorem, it is given below and in the illustration overleaf.

Let ABC be a right-angled triangle with the right angle at A, so that BC is the hypotenuse. Join F to C and A to D, and drop a perpendicular from A to the line DE, meeting it at H and cutting BC at J.

Now, in the two triangles ABD and FBC,

$$AB = FB \text{ and } BC = BD$$
$$\angle FBC = \angle ABC + 90°$$
$$\angle ABD = \angle ABC + 90°.$$

Triangles with two sides and the included angle equal are equal in all respects. So $\triangle ABD$ and $\triangle FBC$ are equal in area.

But the area of any triangle is equal to half the area of a rectangle on the same base and with the same altitude. So that

$$\triangle FBC = \tfrac{1}{2} \text{ rectangle FGAB}$$

and $$\triangle ABD = \tfrac{1}{2} \text{ rectangle BJHD.}$$

Thus these two rectangles are equal, and the square FGAB is equal in area to the rectangle BJHD.

Clearly, with a similar construction on the right-hand side of the triangle ABC we can also prove the square on AC to be equal in area to the rectangle JCEH; and the sum of the two rectangles is BCED, the square on BC.

If the implications of Pythagoras's theorem were confined to right-angled triangles its value would not be very great; but plainly that is not so. Any triangle is, as is clearly seen, the sum of two right-angled triangles. Furthermore, if the respective angles of two triangles are equal, the ratios of their respective sides will also be equal.

(Below and opposite) a step-by-step proof of Pythagoras theorem

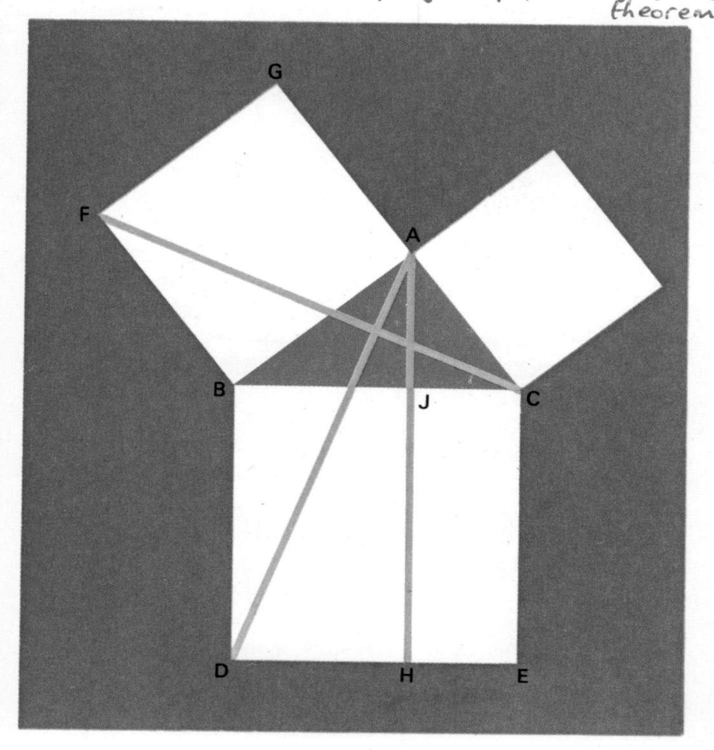

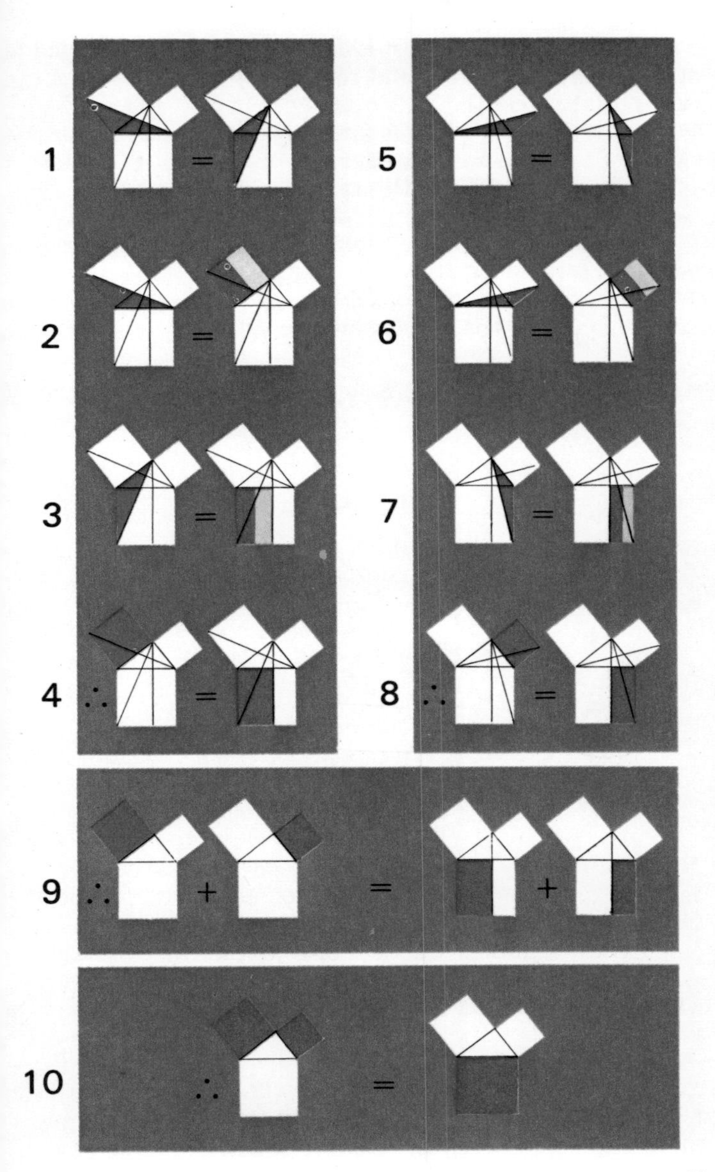
1
=
2
=
3
=
4
=
5
=
6
=
7
=
8
=
9
+
=
+
10
=

Given that the angle at A is equal to the angle at E, that at B equal to that at F, and so on, it is clear that all the ratios will be equal, *i.e.*, AB : AC :: EF : EG; AD : AC :: EH : EG, and so on. Thus, if the angle at B (and of course that at F) is known, then the ratio of the side AD to the side AB (and of EH to EF) is also constant and can be calculated.

These are known as the trigonometrical ratios; and it is advisable to memorize them:

$$\frac{AD}{AB} = \text{sine B (abbreviated to sin B)}$$

$$\frac{BD}{AB} = \text{cosine B (abbreviated to cos B)}$$

$$\frac{AD}{BD} = \text{tangent B (abbreviated to tan B).}$$

There is a secondary group of trigonometrical ratios, the secant, cosecant and cotangent, which are reciprocals of the sine, cosine and tangent respectively. Their ratios are as follows:

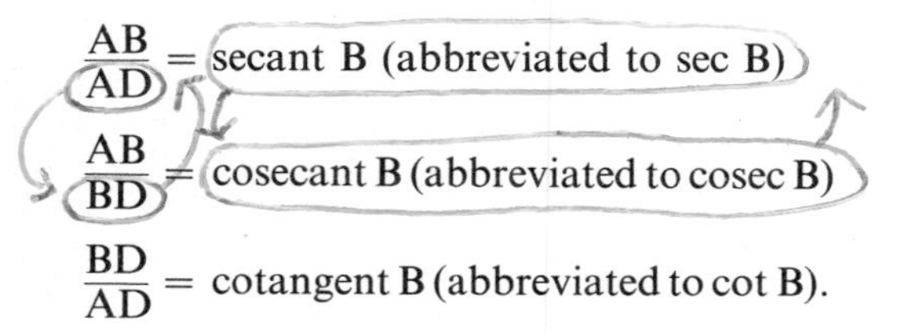

$$\frac{AB}{AD} = \text{secant B (abbreviated to sec B)}$$

$$\frac{AB}{BD} = \text{cosecant B (abbreviated to cosec B)}$$

$$\frac{BD}{AD} = \text{cotangent B (abbreviated to cot B).}$$

Now these ratios, called the reciprocal ratios, can sometimes be used to solve problems where to use the principal functions (the sine, cosine and tangent) would involve us in lengthy calculation. All we need to remember is that

$$\sec B = \frac{1}{\sin B}, \quad \operatorname{cosec} B = \frac{1}{\cos B}, \quad \text{and } \cot B = \frac{1}{\tan B}.$$

However, some ratios are very easily worked out. If, for example, the right-angled triangle ABC has the angle at C a right angle and if the angle at B is 30°, we can draw underneath it the 'mirror' triangle DBC. Now the angle

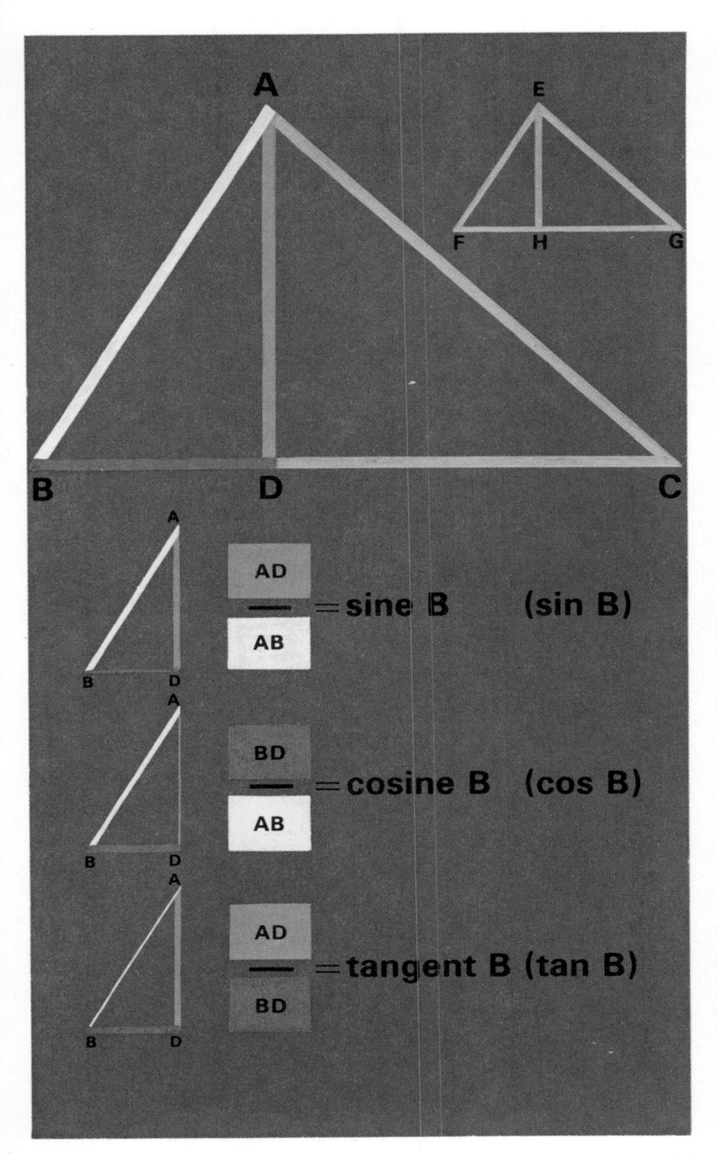
A
E
F
H
G
B
D
C
A
B
D
AD
AB
= sine B (sin B)
A
B
D
BD
AB
= cosine B (cos B)
A
B
D
AD
BD
= tangent B (tan B)

at A is 60° (because the three angles of any triangle are together equal to two right angles) and so is the angle at D; and $\angle ABD$ is obviously also 60°. Thus the triangle ABD is equilateral, so that AB = AD = 2AC, and the ratio of AC to AB, i.e. sin B, is $\frac{1}{2}$.

It is not difficult but very tedious to work out the trigonometrical ratios for all angles; fortunately the job has already been done for you – you have only to consult the tables of sines, cosines and tangents or secants, cosecants and cotangents. By using these you can calculate all the proportions of *any* triangle, right-angled or not, provided you have the necessary minimum of information.

We can *construct* a triangle if we are in possession of any one of the following three sets of information:

For sin 8° 25′, find 8 in the degree column, look along the row to the minute column headed 24′ and to that number add 3, the figure in the mean difference column headed 1′. So sin 8° 25′ = ·1464.

NATURAL SINES

	0′	6′	12′	18′	24′	30′	36′	42′	48′	54′	1′	2′	3′	4′	5′
0°	·0000	0017	0035	0052	0070	0087	0105	0122	0140	0157	3	6	9	12	15
1	·0175	0192	0209	0227	0244	0262	0279	0297	0314	0332	3	6	9	12	15
2	·0349	0366	0384	0401	0419	0436	0454	0471	0488	0506	3	6	9	12	15
3	·0523	0541	0558	0576	0593	0610	0628	0645	0663	0680	3	6	9	12	15
4	·0698	0715	0732	0750	0767	0785	0802	0819	0837	0854	3	6	9	12	14
5	·0872	0889	0906	0924	0941	0958	0976	0993	1011	1028	3	6	9	12	14
6	·1045	1063	1080	1097	1115	1132	1149	1167	1184	1201	3	6	9	12	14
7	·1219	1236	1253	1271	1288	1305	1323	1340	1357	1374	3	6	9	12	14
8	·1392	1409	1426	1444	1461	1478	1495	1513	1530	1547	3	6	9	12	14
9	·1564	1582	1599	1616	1633	1650	1668	1685	1702	1719	3	6	9	11	14
10	·1736	1754	1771	1788	1805	1822	1840	1857	1874	1891	3	6	9	11	14
11	·1908	1925	1942	1959	1977	1994	2011	2028	2045	2062	3	6	9	11	14
12	·2079	2096	2113	2130	2147	2164	2181	2198	2215	2233	3	6	9	11	14
13	·2250	2267	2284	2300	2317	2334	2351	2368	2385	2402	3	6	8	11	14
14	·2419	2436	2453	2470	2487	2504	2521	2538	2554	2571	3	6	8	11	14
15	·2588	2605	2622	2639	2656	2672	2689	2706	2723	2740	3	6	8	11	14
16	·2756	2773	2790	2807	2823	2840	2857	2874	2890	2907	3	6	8	11	14
17	·2924	2940	2957	2974	2990	3007	3024	3040	3057	3074	3	6	8	11	14
18	·3090	3107	3123	3140	3156	3173	3190	3206	3223	3239	3	6	8	11	14
19	·3256	3272	3289	3305	3322	3338	3355	3371	3387	3404	3	5	8	11	14
20	·3420	3437	3453	3469	3486	3502	3518	3535	3551	3567	3	5	8	11	14

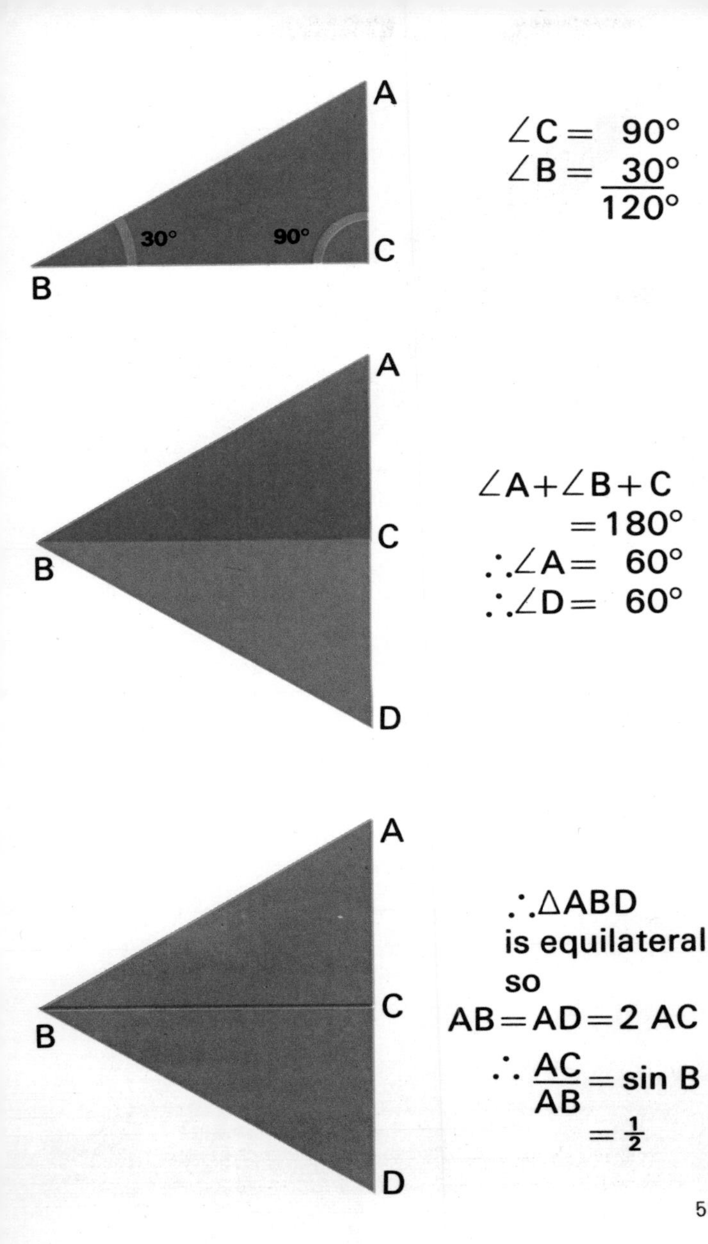
A
30°
90°
C
B
∠C = 90°
∠B = 30°
120°
A
C
B
D
∠A + ∠B + C
= 180°
∴∠A = 60°
∴∠D = 60°
A
C
B
D
∴ΔABD
is equilateral
so
AB = AD = 2 AC
∴ AC/AB = sin B
= 1/2

A triangle
can be constructed
from any one of these
sets of information

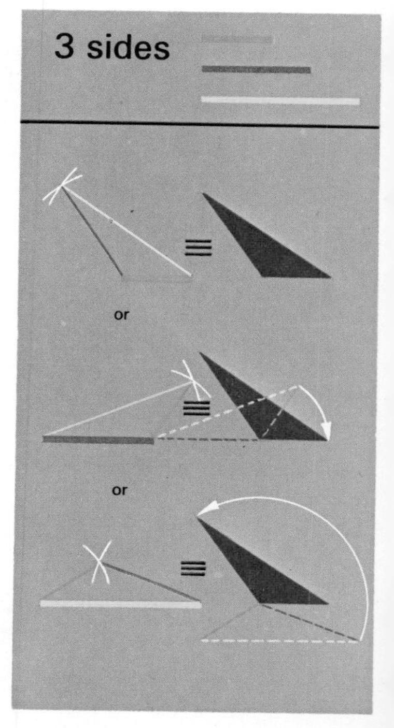

(a) the respective lengths of the three sides of the triangle; (b) the lengths of any two sides and the included angle; or (c) the length of one side and any two angles. The same information therefore suffices to calculate its area and its other properties. The formulae for finding the area are so useful as to be worth remembering. They are:*

(a) when the respective lengths of the three sides are known, the area equals $\sqrt{s(s-a)(s-b)(s-c)}$,

(b) when the side a and the two angles at B and C are known, the area will be $\dfrac{a^2 \tan B \tan C}{2(\tan B + \tan C)}$,

(c) when the sides a and c and the angle at B are known, the area will be $\dfrac{ac \sin B}{2}$.

* We use a, b and c to represent the lengths of the respective sides, a being the side opposite to the angle A, and so on; s stands for half the perimeter (the sum of the three sides), *i.e.*, $s = \dfrac{a+b+c}{2}$.

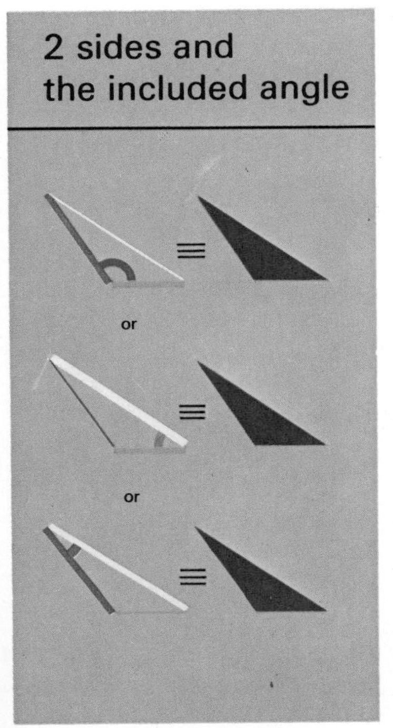

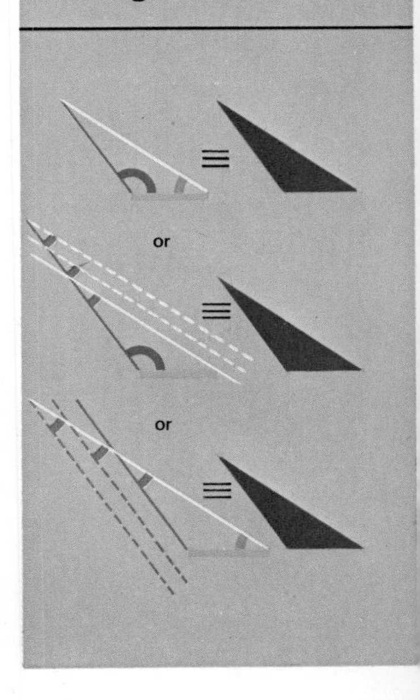

All this, the basis of trigonometry, is directly derived from, and can be proved by, the theorem of Pythagoras; those who denigrate Euclid should pause and consider how surveyors would cope with their work without his aid.

Supposing it is desired to find the area of an irregular five-sided estate, the surveyor can produce his estimate without carrying out more than *one* actual measurement. Suppose each of the five corners to be marked by flags (or possibly there may be some natural features of the landscape to distinguish them). The surveyor with his measuring tape lays out *one* straight line of any length he chooses, and then from each end of this line he measures with his theodolite the angle made with it by a line drawn to each of the five corners. He does this by levelling up the instrument, lining up the telescopic viewer on the appropriate flag or natural feature and reading off the angle on the horizontal scale. Now in all these triangles CAB, DAB, EAB and so on he knows the length of the line AB (because he has measured it out) and two angles. Hence he can arrive at all the measurements of these triangles and, by extension, at all the measurements he needs.

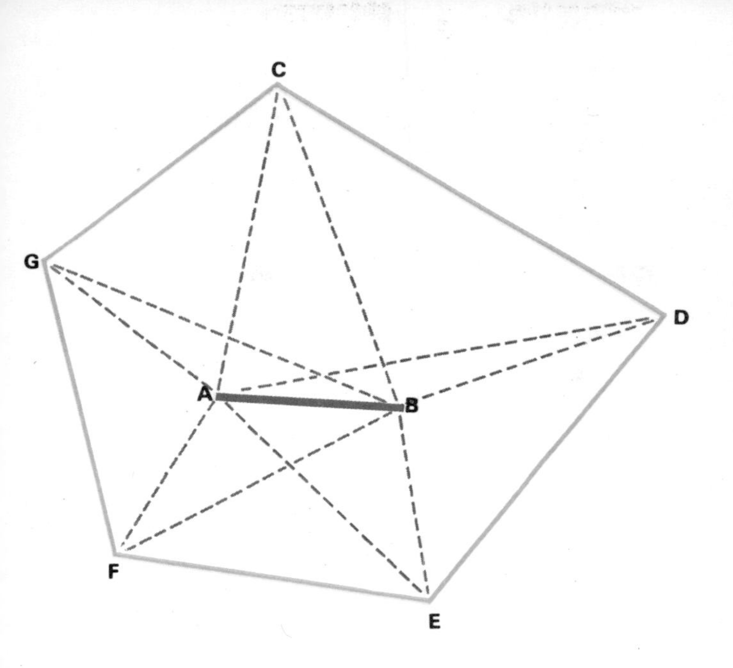

A surveyor can calculate the area of an irregular five-sided estate by measuring *one* straight line of any length. From each end of the line he measures the angle made with it by a line drawn to each of the five corners. This gives him the length of the line AB and two angles in each triangle, from which he can calculate all required measurements.

ALL ABOUT NOTHING

'. . . gives to airy nothing
A local habitation and a name' – *Shakespeare*

In our system of numeration we are so used to the symbol '0' that we may be tempted to suppose it 'grew naturally' and was always there. But this is by no means the case. To go back no further than a couple of thousand years, the Romans had no symbol for nought, and their calculations were, largely because of that, incredibly complicated. None of us would have any difficulty in multiplying, say, 44 by 187. But imagine coping with the task of multiplying XLIV by CLXXXVII !

I must confess I do not know how the Romans did their multiplication (or, for that matter, the English who up to the thirteenth century were still using Roman numerals). I can only suppose it was accomplished step by step by some such formidable means as those shown opposite.

The symbol for nought was introduced to the world by the Indians; and it is at the same time one of the most useful and one of the most dangerous numbers that have ever been invented. Dangerous partly because of its ambiguities.

XLIV × CLXXXVII

XL × C	MMMM			
XL × L	MM			
XL × X		CCCC		
XL × X		CCCC		
XL × X		CCCC		
XL × V		CC		
XL × II			LXXX	
IV × C		CCCC		
IV × L		CC		
IV × XXX		C	XX	
IV × V			XX	
IV × II				VIII

MMMMMMCCCCCCCCCC, CCCCCCCCCC, CLXXXXX, XXVIII

= MMMMMMMMCLXXXXX, XXVIII

= MMMMMMMMCCXXVIII

× First of all, like all other integers, nought has more than one meaning. The symbol '2', for example, means something quite different in the number 324 from what it means in the number 572. That is its positional value as distinct from its absolute value. While 0 has no numerical value its positional value is just as important as that of any other number. In the two numbers 4,302 and 4,032 the position of the 0 has made a difference of 270 – a difference which the Romans found it difficult to express in their system of numeration.

Then again, as stated elsewhere, every number other than 0 and infinity (with which we shall be concerned in the next section) can be used either as number *qua* number or as an operative. In $7 - 3$, seven is the number and three the operative (*from* 7 *take* 3). It may seem to make little difference whether you write 0×7 (multiply 0 *by* 7) or 7×0 (multiply 7 *by* 0), but it is in fact better to use the former construction. If we never use 0 or infinity as operatives we shall escape a number of difficulties. Dividing by 0 ($x \div 0 =$ infinity) is liable to cause us to fall into quite a number of traps.

Probably nobody would be so silly as to suppose that because $0 \times 6 = 0$ and $0 \times 7 = 0$ therefore $6 = 7$. But dress this up a little and we come to a well-known fallacy that has baffled many a schoolboy.

To prove that $1 = 2$.

Let $$x = y = 1.$$
Then $$x^2 - xy = x^2 - y^2.$$
So $$x(x - y) = (x + y)(x - y).$$
Dividing both sides by $x - y$,
$$x = x + y,$$
and $$1 = 1 + 1 = 2.$$

But $x - y = 0$, so what we have done here is to divide each side of the equation by 0. If we divide any number by 0 (a procedure not well seen by most mathematicians) the answer – presuming there to be an answer at all – is infinity; so the only (and not very useful) conclusion we have arrived at is that infinity is equal to infinity.

There is another danger inherent in the use of 0 to mean not zero but a quantity too small to be worth bothering about. For instance, we know that the fraction $\frac{1}{3}$ can be expressed in decimal notation as $\cdot\dot{3}$, or $\cdot 3333\ldots$, so that we say that $\frac{1}{3}$ is 'equal' to $\cdot\dot{3}$. So it is, nearly enough for most practical purposes. But if we look more carefully we shall see that $\cdot 3 = \frac{3}{10}$, which is less than $\cdot 4$; $\cdot 33 = \frac{33}{100}$, which is less than $\cdot 34$; $\cdot 333 = \frac{333}{1000}$, which is less than $\cdot 334$, and so on, so that $\frac{1}{3}$ is greater than $\cdot 3333 \ldots 3$ and less than $\cdot 3333 \ldots 4$. Since we can go on with these decimal figures as long as we like, we are justified in saying that $\cdot\dot{3}$ is *in practice* equivalent to $\frac{1}{3}$; it is more accurate and safer to say that $\cdot\dot{3}$ *when indefinitely extended approaches* the value $\frac{1}{3}$. We shall come back to this point when we deal with series.

The symbol for nought, which was introduced by the Indians, adopted by the Arabs and transmitted to Europe via Moorish culture, stood for the empty column of the abacus, making it possible to represent any number with the aid of only nine other symbols.

HOW MUCH IS INFINITY?

'What you see, yet cannot see over, is as good as infinite' – Thomas Carlyle
'Ah, but a man's reach should exceed his grasp,
Or what's heaven for?' – Robert Browning

If you were to start to write down all the natural numbers (1, 2, 3 ... and so on) and to continue this not very enthralling occupation all your life, you would obviously attain to a very high number indeed. But you would never reach the end of the series, simply because it *has* no end. So mathematicians say that the imagined last term in this series would be infinity (the sign for which is ∞). It is a concept that has to be employed in some branches of mathematics; but its use can be very tricky, since it does not obey all the laws that govern finite numbers.

That you cannot add anything to infinity or multiply infinity by another number is not so difficult to realize. You can't have anything better than the best or greater than the greatest. But it is also impossible by any ordinary mathematical process to *diminish* infinity. If out of a sum of a million pounds you spend a penny you have not seriously impoverished yourself, but you have reduced your capital to a measurable extent. Not so with infinity. In fact, where x represents any finite number,

$$\infty + x = \infty$$
$$\infty - x = \infty$$
$$\infty \times x = \infty$$
$$\infty \div x = \infty.$$

Here is a vital difference between 0 and infinity. You cannot increase infinity by any mathematical process;

The nature of infinity expressed by the boy's image reflected an infinite number of times in two facing mirrors

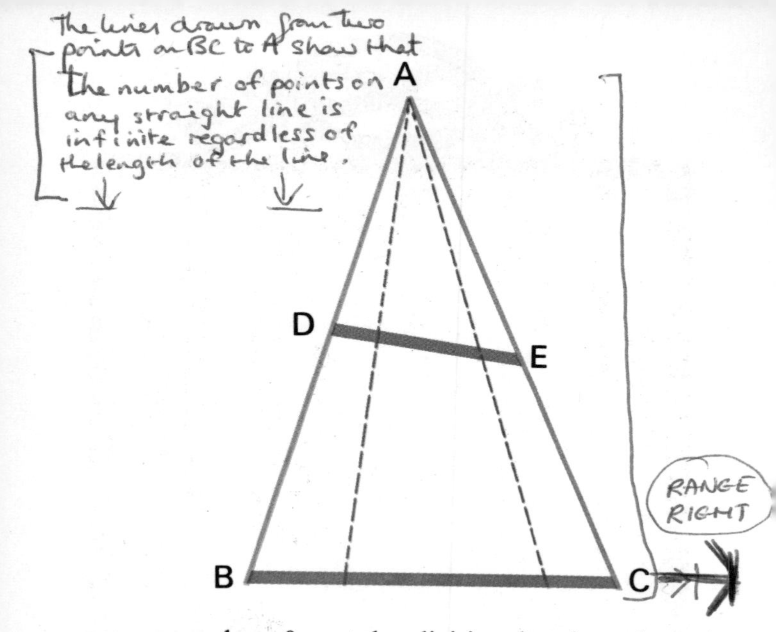

you *can* reduce 0, not by division but by subtraction: $0 - 5 = -5$. In other words, if you have no money in the bank and you cash a cheque for five pounds, you then have minus five pounds (less than nothing) in your account.

If you try to treat infinity as a natural number you run into paradoxical situations, some of which we discuss in the section on 'Sets and Series'.

The number of points on a straight line is infinite, no matter what the length of the line may be. It is a little difficult to realize that there are just as many points on a line measuring one inch as on a line measuring two inches; but this is quite easily seen to be the case.

Let us suppose, see the example above, that we draw a line of two inches and another one above it measuring one inch, join the ends together and produce these two lines until they meet, forming a triangle. Now if we choose any point on the line BC and join it by a straight line to A, that line must cut DE at some point or other, so that for every point on the two-inch line BC there must be a corresponding point on the one-inch line DE. But we must avoid the trap of saying that because each line contains the same number of points, therefore one inch is equal to two inches.

The sum of an infinite series is not necessarily infinitely great. This too will become apparent when we study series.

To take an example in which we have to deal with both 0 and infinity. Suppose we are given the equation $xy = 36$ (this, as we shall see in the section 'Mathematics in Pictures', is an equation which if plotted on a graph produces the curve known as a hyperbola). Now the successive values for x and y (considering for the moment only positive integers) are as given in the chart below.

By taking fractional values we can continue to do this as long as we like, when we shall find that as x increases to an enormous sum (say, 36 millions) y becomes very tiny (one-millionth), and it is correct to state that in this equation as x (or y) *tends* to approach infinity y (or x) *tends* to approach 0. But neither of them ever reaches the goal, so that we are saved from having to cope with the meaningless statement $0 \times \infty = 36$.

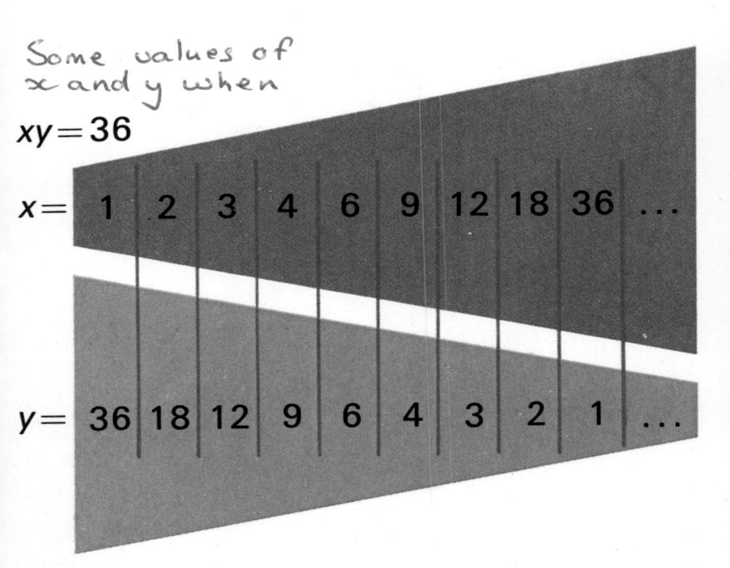

1 (2^0) 2 (2^1) 4 (2^2)

SETS AND SERIES

'How can finite grasp infinity?' – Dryden

A series is a set of terms each of which stands in a constant mathematical relationship to that which follows it. This relationship may be of addition (or subtraction) of a fixed number or of multiplication (or division) by a fixed number, or the relationship may be more complex.

The simplest series is, of course, the set of 'natural' numbers, 1, 2, 3, etc., arranged in ascending order, where the relationship is simply that each term is one greater than the term that precedes it. However far this series may extend, it is obvious that the nth term will be n. But what will be the sum of the series to n terms?

The sum will always be $\frac{n(n+1)}{2}$, as is very easily seen:

$$S = 1 + 2 + 3 + \ldots + (n-2) + (n-1) + n$$

and $S = n + (n-1) + (n-2) + \ldots + 3 \ldots + 2 + 1.$

Adding these together, it is plain that

$$2S = (1+n) + (2+n-1) + (3+n-2) + \ldots + (n-2+3) + (n-1+2) + (n-1)$$

$$= (n+1) + (n+1) + (n+1) + \ldots n \text{ times},$$

so that $$2S = n(n+1)$$

and $$S = \frac{n(n+1)}{2}.$$

Trying this out with actual figures,

$$1 + 2 + 3 + 4 + 5 + 6 = 21$$

and $$\frac{6.7}{2} = 21.$$

A series that increases (or decreases) by a constant difference is called an Arithmetical Progression. Of course, it need not begin with unity, nor need the difference each time be one. A general formula to sum any Arithmetical

Progression is:*

$$S = \frac{n}{2}\{2a + (n - 1)d\}.$$

Try this with any Arithmetical Progression you like, *e.g.*, $5 + 11 + 17 + 23 + 29 + 35 + 41 = 161$. Here we have $n = 7$, $a = +5$, $d = +6$, and

$$\frac{n}{2}\{2a + (n - 1)d\} = \frac{7}{2}\{10 + (6.6)\}$$

$$= \frac{7.46}{2} = 7.23 = 161.$$

Or a diminishing progression:

$$+8 + 3 - 2 - 7 - 12 - 17 = -27.$$

Here $n = 6$, $a = +8$, $d = -5$, and

$$\frac{n}{2}\{2a + (n - 1)d\} = \frac{6}{2}\{+16 + 5(-5)\}$$

$$= 3(16 - 25) = 3(-9) = -27.$$

A Geometrical Progression can be summed by a similar procedure:

$$S = a + ar + ar^2 + \ldots ar^{n-2} + ar^{n-1} \ldots$$

Multiplying this by r, we get:

$$rS = ar + ar^2 + ar^3 + \ldots ar^{n-1} + ar^n$$

and $S = a + ar + ar^2 + \ldots ar^{n-2} + ar^{n-1}$.

If the second line be subtracted from the first, all the intermediate terms cancel out, leaving:

$$rS - S = ar^n - a, \text{ or } S = \frac{a(r^n - 1)}{r - 1}.$$

Try this out with the series $2 + 6 + 18 + 54 + 162 = 242$, giving $n = 5$, $a = +2$, $r = +3$:

$$\frac{a(r^n - 1)}{r - 1} = \frac{2(3^5 - 1)}{3 - 1} = \frac{2.242}{2} = 242.$$

* In dealing with series it is convenient to employ a number of abbreviations: S stands for the sum of the series, n for the number of terms in it, a for the first term, d for the constant difference (if that difference is one of addition or subtraction), and r for the ratio (if the difference is one of multiplication or division).

In dealing with more complicated series we meet the figurate numbers, the first of them being the triangular numbers, derived from the natural numbers by taking first the original first term, then the sum of the first two terms, then the sum of the first three, and so on:
1, (1 + 2), (1 + 2 + 3), (1 + 2 + 3 + 4), ... or 1, 3, 6, 10 ...

Why are they called 'triangular' numbers? Because in each case, if you picture them by representing each unit as a dot, the dots form a triangle.* Similarly the square numbers can be shown as squares of dots, and so on.

The triangular numbers can also be written as:

$\frac{1.2}{2}, \frac{2.3}{2}, \frac{3.4}{2}$, and so on, so that the $(n - 1)$th term will be $\frac{n(n-1)}{2}$ and the nth $\frac{n(n+1)}{2}$.

Any square is the sum of two successive triangular numbers; see the illustration, or if you prefer an algebraic proof:

$$\frac{n(n-1)}{2} + \frac{n(n+1)}{2} = \frac{n(n-1+n+1)}{2} = \frac{2n^2}{2} = n^2.$$

To sum these and more complicated series, excluding Geometrical Progressions and their derivatives, use the Vanishing Triangles. You write the series, and then in successive lines the differences (plus or minus), then the differences between those differences, and so on until you reach a line of 0's. So, using the triangular numbers:

```
1         3         6         10         15
     +2        +3        +4         +5
          +1        +1         +1
               0          0
```

* They don't, of course, in the case of the number one. But unity is regarded as an 'honorary member' of the family of figurate numbers.

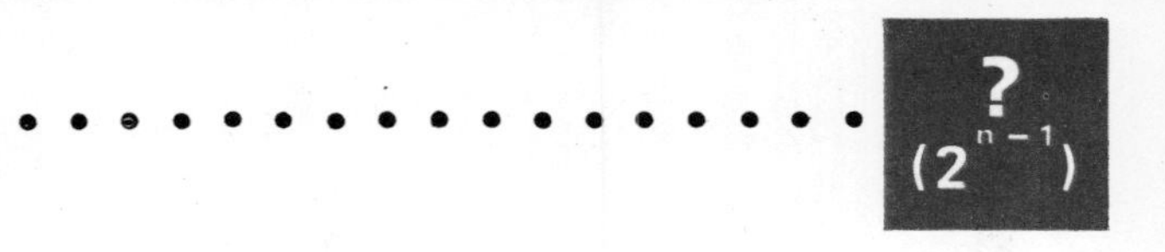

To sum the series, add together the left-hand figure of the first line multiplied by $\frac{n}{1}$, the left-hand figure of the second line multiplied by $\frac{n(n-1)}{1.2}$, the left-hand figure of the third line by $\frac{n(n-1)(n-2)}{1.2.3}$, and so on:

$$n + \frac{2n(n-1)}{1.2} + \frac{n(n-1)(n-2)}{1.2.3}$$

$$= n + (n^2 - n) + \frac{n^3 - 3n^2 + 2n}{6}$$

$$= \frac{6n^2 + (n^3 - 3n^2 + 2n)}{6} = \frac{n^3 + 3n^2 + 2n}{6}$$

$$= \frac{n(n+1)(n+2)}{6}.$$

Try this out:

$$1 + 3 + 6 + 10 + 15 + 21 = 56.$$

Using the same formula: $\frac{6.7.8}{6} = 56.$

Deal with the square numbers in the same way:

$$\begin{array}{ccccccccc} 1 & & 4 & & 9 & & 16 & & 25 \\ & +3 & & +5 & & +7 & & +9 & \\ & & +2 & & +2 & & +2 & & \\ & & & 0 & & 0 & & & \end{array}$$

$$S = \frac{n}{1} + \frac{3n(n-1)}{1.2} + \frac{2n(n-1)(n-2)}{1.2.3}$$

$$= \frac{6n}{6} + \frac{9n^2 - 9n}{6} + \frac{2n^3 - 6n^2 + 4n}{6}$$

$$= \frac{2n^3 + 3n^2 + n}{6} = \frac{n(n+1)(2n+1)}{6}.$$

Try this out:

$$1 + 4 + 9 + 16 + 25 = 55.$$

Using the formula: $\frac{5.6.11}{6} = 55.$

To find the nth term in the series, we add together the left-hand figure in the first line, the left-hand figure in the second line multiplied by $\frac{n-1}{1}$, the left-hand figure of the third line by $\frac{(n-1)(n-2)}{1.2}$, and so on.

Try this with the cubes:

$$\begin{array}{ccccccccc} 1 & & 8 & & 27 & & 64 & & 125 \\ & +7 & & +19 & & +37 & & +61 & \\ & & +12 & & +18 & & +24 & & \\ & & & +6 & & +6 & & & \\ & & & & 0 & & & & \end{array}$$

Then the fifth term (125) will be:

$$1 + 7(5-1) + \frac{12(5-1)(5-2)}{1.2} + \frac{6(5-1)(5-2)(5-3)}{1.2.3} = 1 + 28 + 72 + 24 = 125$$

and the sum will be:

$$S = \frac{1.n}{1} + \frac{7n(n-1)}{1.2} + \frac{12n(n-1)(n-2)}{1.2.3} + \frac{6n(n-1)(n-2)(n-3)}{1.2.3.4} = \tfrac{1}{4}(4n + 14n^2 - 14n + 8n^3 - 24n^2 + 16n + n^4 - 6n^3 + 11n^2 - 6n)$$

$$= \tfrac{1}{4}(n^4 + 2n^3 + n^2)$$

$$= \frac{n^2(n+1)^2}{4} = \left\{\frac{n(n+1)}{2}\right\}^2.$$

The sum of a series of cubes from unity is always a square.

Certain series (those known as convergent) can be summed to infinity; these are important in mathematics, since the values of some transcendental numbers, such as π and e, can be found by summing such series. Arithmetical Progressions can never be summed in this way, but decreasing Geometrical Progressions can. Take, for example, the series $1 + \frac{1}{2} + \frac{1}{4} + \frac{1}{8} + \dots$ If you add these terms one by one you arrive at the successive totals 1, $1\frac{1}{2}$, $1\frac{3}{4}$, $1\frac{7}{8}$, etc., from which it is clearly seen that the total, while it increases all the time, can never reach 2. It is therefore convenient to say that the sum taken to infinity 'is' 2. Let us see how this operates with the formula

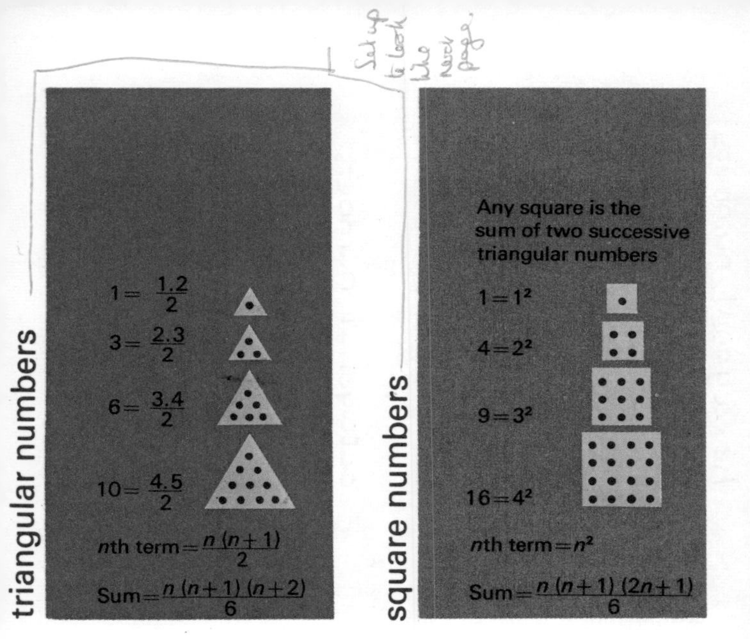

$S = \dfrac{a(r^n - 1)}{r - 1}$, n being infinite, a being 1, and r being $\frac{1}{2}$.
Any fraction less than one multiplied by itself many times becomes a very small quantity, so we say $\frac{1}{2}$ to the power of infinity is 0. Thus the formula becomes

$$\frac{1(0 - 1)}{\frac{1}{2} - 1} = \frac{-1}{-\frac{1}{2}} = 2.$$

Note that if a series grows with increasingly small differences, that does not necessarily make it convergent. Take the Harmonic Progression $1 + \frac{1}{2} + \frac{1}{3} + \frac{1}{4} + \frac{1}{5} + \ldots$ Although the increases become progressively smaller, the sum of the series, taken far enough, will be infinitely great. This is apparent if the series be written as:

$$1 + \tfrac{1}{2} + (\tfrac{1}{3} + \tfrac{1}{4}) + (\tfrac{1}{5} + \tfrac{1}{6} + \tfrac{1}{7} + \tfrac{1}{8}) + \ldots$$

The term $(\frac{1}{3} + \frac{1}{4})$ has two fractions the smaller of which is $\frac{1}{4}$, so that its value exceeds $\frac{1}{2}$; the term $(\frac{1}{5} + \frac{1}{6} + \frac{1}{7} + \frac{1}{8})$ has four fractions of which the smallest is $\frac{1}{8}$, so this value also exceeds $\frac{1}{2}$; and so on. Thus, $1 + \frac{1}{2} + \frac{1}{3} + \frac{1}{4} + \frac{1}{5} + \ldots$ exceeds $1 + \frac{1}{2} + \frac{1}{2} + \frac{1}{2} + \ldots$, and its sum is infinitely great.

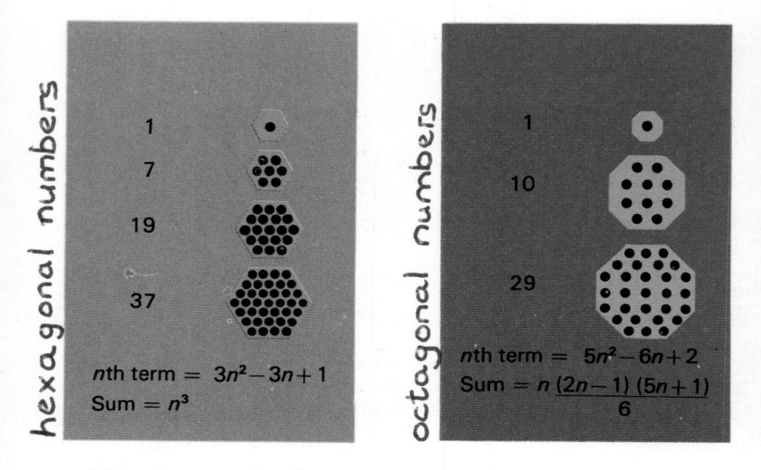

The theory of sets is of great importance in modern mathematics, although a 'set' is not in itself merely a mathematical concept. A set is simply a collection of entities all of which have something in common, and many of them, of course, include less numerous sub-sets. The set of all human beings, for example, includes the sub-set of all males, which again includes the sub-sub-set of all Englishmen, and so on. And of course one entity may be a member of many non-inclusive sets. I am, for example, a member of the set of males and also of that of adults; my wife is a co-member of the latter set but not of the former.

It is the infinite sets in mathematics that have been the most fruitful in number theory and that are also, it must be added, of the greatest difficulty. There are two classes of them, the denumerable and the non-denumerable. An example of the first category is the set of all the natural numbers, 1, 2, 3 and so on. They are called denumerable because they can be labelled (in fact they label themselves) as 1, 2, 3 and so on. The number of points in a line is, however, non-denumerable. It is impossible to number them off 1, 2, 3 and so on because no matter how close members 1 and 2 may be it is always possible to insert another point between them.

Now, the simplest way of counting is by what is called one-to-one correspondence. If you and I know no mathematics at all and we wish to share an even number of apples equally between us, we can do so without bothering with any counting. All that is needed is that you should take an apple each time I take one, and then we shall know that the division has been fairly carried out.

Now let us take the denumerable set of all natural numbers and its included sub-set, the set of all even numbers. First we double each member of the main group, which, while increasing the *value* of each term, has clearly not affected the *number* of terms. They are still denumerable; we can place number 2 first, number 4 second, and so on, attaching each of these doubled numbers to its undoubled parent thus:

$$\begin{array}{cccccc} 1 & 2 & 3 & 4 & 5 & \dots \\ 2 & 4 & 6 & 8 & 10 & \dots \end{array}$$

and we can continue doing so all the way through. There are exactly the same number of terms in the second row as in the first one. But now we find to our dismay that the second line is in fact the sub-set of all the even numbers, and thus that it is co-extensive with a set that includes all its members as well as the equally large sub-set of all the uneven numbers. We are thus forced to the conclusion – however repugnant to common-sense, it is nevertheless inescapable – that an infinite sub-set may contain precisely as many members as does its parent and apparently much more numerous set.

Again, supposing we attempt to sum all the natural numbers, calling their total ∞, double each term in the series, thus of course doubling the total. Now

$$1 + 2 + 3 + 4 + \dots = \infty.$$

So

$$2 + 4 + 6 + 8 + \dots = 2\infty.$$

But the set of all natural numbers includes all the numbers in the second set as well as infinitely many others, so that ∞ is vastly greater than 2∞. We have therefore to accept the conclusion of Georg Cantor, the German mathematician, that there are degrees of infinity, unless we choose to paraphrase George Orwell and say: 'All infinities are equal, but some are more equal than others'.

MATHEMATICS IN PICTURES

'Seeing is believing' – *Old proverb*

Graphs are familiar to everybody, including even those to whom mathematics is a closed book. Constantly one sees in the newspapers graphs showing the increase (rarely, oddly enough, the decrease) in the sales of one or another product. These curves (always provided they are based on accurate information) are not without value. They are readily appreciable by the non-mathematician. If the curve goes constantly upwards, then clearly the product is proving popular. Not only that, but it is easily seen that a curve like this is more encouraging than one like this. In both cases sales are increasing; but in the former case the ratio of increase is also increasing; we are on the up-and-up. In the flattened curve improvement is not being maintained at the same rate; it constitutes a danger-signal.

But graphs are of value in many other respects. First, they provide a simple method of solving equations (though

This sales manager has every reason to feel very pleased with himself. As the graph behind him demonstrates, sales have more than trebled since the product he is selling was introduced and the ratio of increase is still increasing. (Opposite) a sample graph layout demonstrating the statement $x = +3$, $y = -4$.

SALES

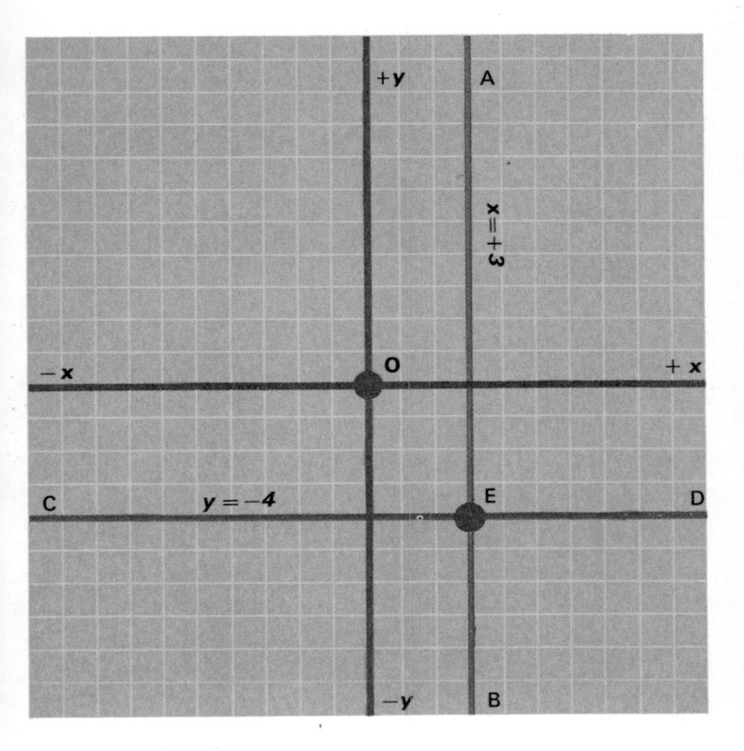

sometimes only approximately). And secondly and more importantly, they will sometimes reveal truths that, although they can be shown by algebra to be valid, are more easily comprehensible if demonstrated graphically. This function will be described at the end of the section.

For those not familiar with graphs it should be explained that the squared sheet of paper is divided into four by two intersecting straight lines, the horizontal one being known as the x axis, the vertical as the y axis. From their point of intersection (conventionally marked O) any plus value of x is indicated by counting x places to the right, any minus value to the left, of the vertical axis; any plus value of y is indicated by counting y places

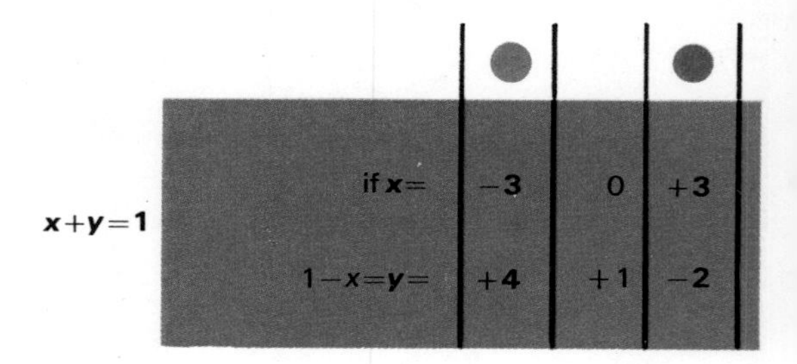

$x+y=1$				
	if $x=$	-3	0	$+3$
	$1-x=y=$	$+4$	$+1$	-2

$x-6y=-27$				
	if $x=$	-9	$+3$	$+9$
	$27+x=6y=$	$+18$	$+30$	$+36$
	$y=$	$+3$	$+5$	$+6$

upwards, any minus value by counting y spaces downward, from the horizontal axis. Thus, '$x = +3$' is denoted by a straight line three places to the *right* of the vertical axis and parallel to it; and '$y = -4$' is shown by a straight line four spaces *below* the horizontal axis and parallel to it.

Now any point in a plane can be indicated by two co-ordinates; that is to say, it must be at the intersection of two straight lines. Thus, in the diagram on page 75 '$x = +3$' is indicated by the straight line AB, '$y = -4$' is indicated by the line CD, and the point at which the two lines cross conveys to us the dual statement '$x = +3$' and '$y = -4$'.

This enables us to find a simple *visual* method of solving equations. Take, for example, the simultaneous equations:

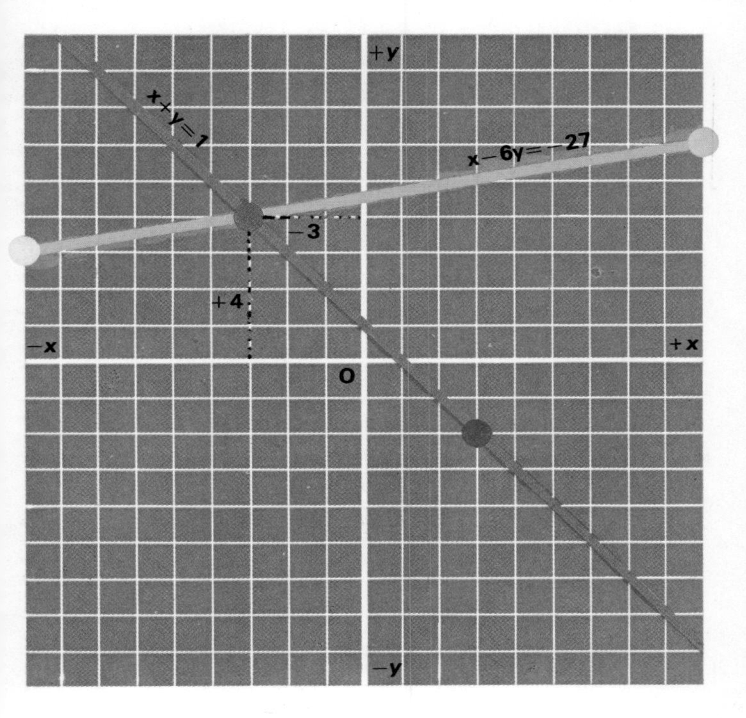

$$x + y = 1$$
$$x - 6y = -27.$$

First we make two short tables showing a few possible values for x and y, as shown opposite.

An equation of the first order (that is to say, one in which x has no index figure higher than 1) is always a straight line; hence it is sufficient to mark only two of the points indicated, join them and produce the line. This will indicate all possible 'real' roots of the equation. We adopt the same procedure with the second equation and draw the straight line indicated by these values; it will intersect the first line at a point three spaces left of centre and four places above centre, and give us a straightforward visual demonstration of the result: $x = -3$, $y = +4$.

Equations of higher orders (which have more than one root) can be solved in the same way, although when x has an index figure higher than 1 the line it traces out will be not a straight line but a curve. Take, for example, the equation $x^2 - 5x = +6$. To construct a graph we must introduce a y co-ordinate, which we do very simply by saying $y = +6$ and turning our equation into $x^2 - 5x = y$.

Here again we construct a table of values, but since we have to draw a curve we shall need more than the two co-ordinates that suffice for a straight line, as illustrated in the table below.

As will be seen from the accompanying diagram, these points indicate a curve, known as a parabola, which always results from an equation of the second order, and all possible values of x are indicated by points on this curve. Now, the second equation ($y = +6$) is indicated by a straight line 6 spaces above the vertical axis, and the solution to the equation is shown by the point or points at which the straight line cuts the curve: in this case $x = +6$ or -1.

Note that we have said 'the point or points'. An equation such as that cited here has two 'real' roots, so that there are two points of intersection. An equation such as $x^2 + 2x = -1 = y$ has no *intersecting* points at all; the straight line does not cut the curve but touches it (in other words it is tangential to the curve), so that there is only

	●	●	●	●	●	●	●
if $x=$	−2	−1	0	+1	+2	+3	+4
$x^2=$	+4	+1	0	+1	+4	+9	+16
$-5x=$	+10	+5	0	−5	−10	−15	−20
$y=$	+14	+6	0	−4	−6	−6	−4

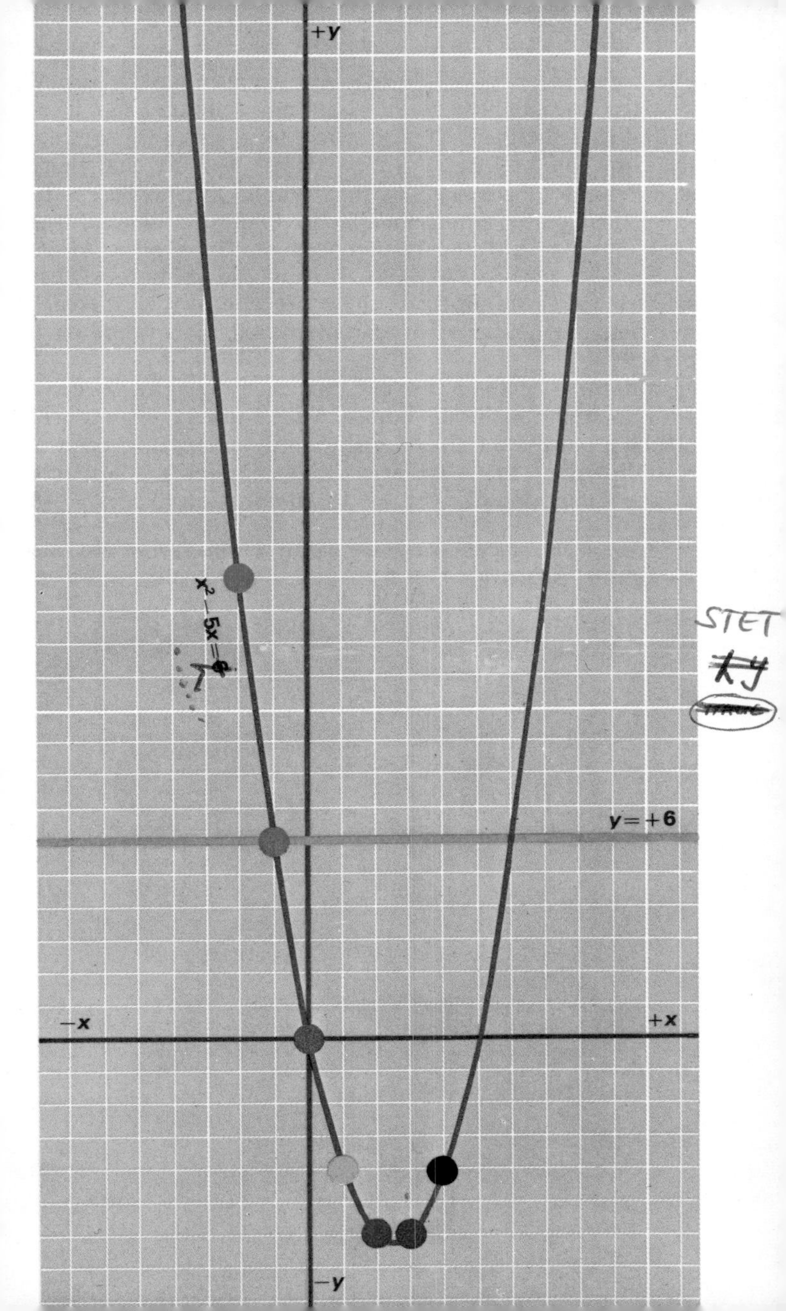
+y
$x^2 - 5x =$
$y = +6$
−x
+x
−y

one solution, $x = -1$.* In some equations it is found that the straight line neither cuts nor touches the curve; in that case there are no 'real' roots to it. Try, for example, the equation $x + 2x = -7 = y$; you will see that there is no point of contact, so that graphs are powerless to solve this sort of equation. By the quadratic formula (see page 30) we find the only possible values of x to be $-1 + \sqrt{-6}$ and $-1 - \sqrt{-6}$ (or, if you prefer to write it so, $-1 \pm i\sqrt{6}$). Both of these are complex (not 'real') numbers, so neither can be represented by a point on a plane.

Equations of the third and higher orders can be dealt with in similar fashion, though in such cases the curve will not be a parabola (which appears only in equations of the second order). In an equation that contains x^3 the curve will be something like this . In an equation like $xy = 10$

* Some mathematicians like to say that *all* quadratics have two solutions, so that in an equation like $x + 2x = -1$ there are the *two* solutions $x = -1$ and $x = -1$, although they are in fact identical.

there will be a double curve (a hyperbola). Equations of the first order have only one solution, of the second order two, and, generally, equations of the *n*th order have *n* solutions, any or all of which may be 'real', imaginary or complex.

More important than the solving of equations, which often yield more easily to other methods, is the utility of graphs in illustrating physical phenomena. An example of this is seen in the first graph we described – what might be called a 'commercial' graph. It can also be used to clarify the comparison of other phenomena, as, for example, weight-for-age. Here you could indicate a baby's age in months by your *x* co-ordinate and its weight in pounds by your *y* co-ordinate (or *vice versa*); the resulting curve would provide visual evidence of the child's progress. You can do this with *any* two phenomena provided (and this, as we shall see when we talk about statistics, is vital) there is some causal connection between them.

Incidentally, graphs provide a means of dealing with that puzzling old problem posed by Zeno of Elea about Achilles and the tortoise. You will doubtless recall it: Achilles pursuing the tortoise can never catch it, because when he has covered the distance that first separated them the tortoise will have progressed further; when he has

Achilles versus the tortoise was a stirring race with which Zeno of Elea bewildered the classical world. Zeno argued that Achilles would never catch the tortoise because each time he covered the distance separating them the tortoise would have progressed further. Our graph overleaf demonstrates the distance Achilles had to run and the time it took him to catch and overhaul the tortoise.

made up that gap the tortoise will have gone some distance further still, and so the tortoise will always maintain its advantage. With graphs we can not merely prove but *show* that Achilles *will* capture his prey, and how far he will have to run in order to do so.

In dealing with a problem by graphical methods it is often convenient to change the scale (which makes no difference so long as we remember to change it back again afterwards). Here we use one small division to mark one second, but in showing distances use a scale of 50 yards to one large division.

Now we prepare our graph, shown opposite, marking the starting point of Achilles at the point O and that of the tortoise at T, 200 yards (four large divisions) to the right along the x axis; we use the y axis to indicate time taken at the rate of one space to the second. The tortoise will travel 5 yards (one small division) in 50 seconds, its progress being indicated by the straight line originating at T: Achilles runs ten yards (two small divisions) in one second, 20 yards in two seconds, and so on, his progress being shown by the straight line originating at O.

Since both courses are represented by straight lines, we need to locate only one point each in addition to the starting points: in 5 seconds Achilles will have reached point A_1 and the tortoise in 50 seconds will have reached point T_1. We join O to A_1 and T to T_1, and produce the A line until it meets the T line at the goal (G). Now if from G we drop perpendiculars to both axes we shall find the distance Achilles has to run and the time it takes him: the perpendicular from G to y axis shows 202 yards approximately, and that from G to the x axis shows 20·2 seconds approximately.

The result can be checked by algebra: if Achilles catches the tortoise in x seconds he will have run $10x$ yards and the tortoise will have crawled $\frac{x}{10}$ yards. So $10x - \frac{x}{10} = 200$, or $99x = 2{,}000$, so $x = 20{\cdot}2$ seconds approximately and the distance run by Achilles will be 202 yards approximately.

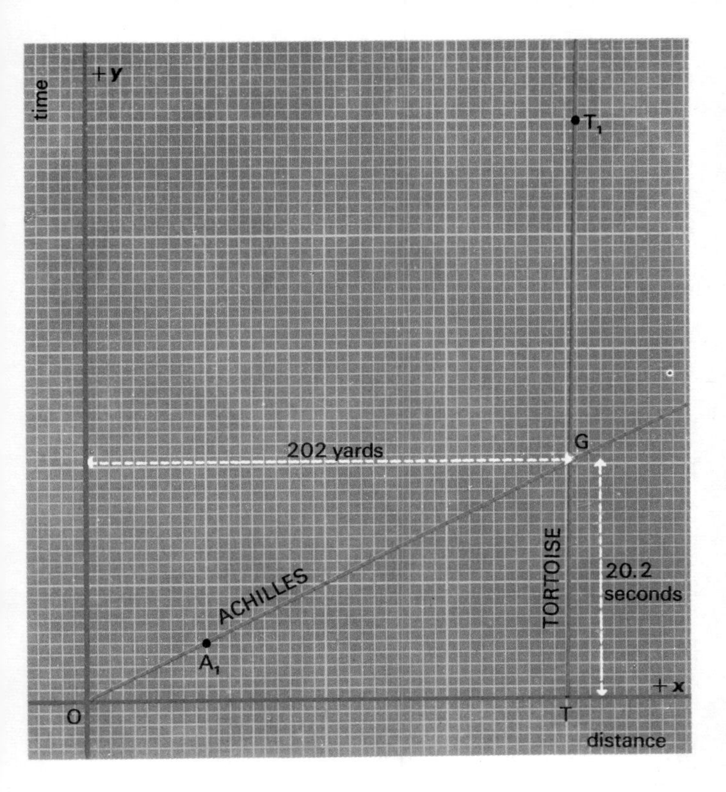

The greatest value of Descartes' Analytical Geometry (as graphs are also called) is the way in which it illustrates physical phenomena based on the conic sections and other curves. For example, both the orbit of a satellite round the earth and a planet round the sun are ellipses and the path of a projectile or that of an object thrown into the air, a cannon ball or a cricket ball, is derived from an inverted parabola, which can be illustrated thus: ⌒.

All of these curves derive from equations. For instance, $x^2 + y^2 = z^2$, the Pythagorean equation of the right-angled triangle, is almost identical with the simplified version of the Equation of the Circle, $x^2 + y^2 = r$. Why?

Take the specific equation $x^2 + y^2 = 625$ and work out the usual table of possible values for x and y, as shown below. Plot this curve and you will have traced out a circle.

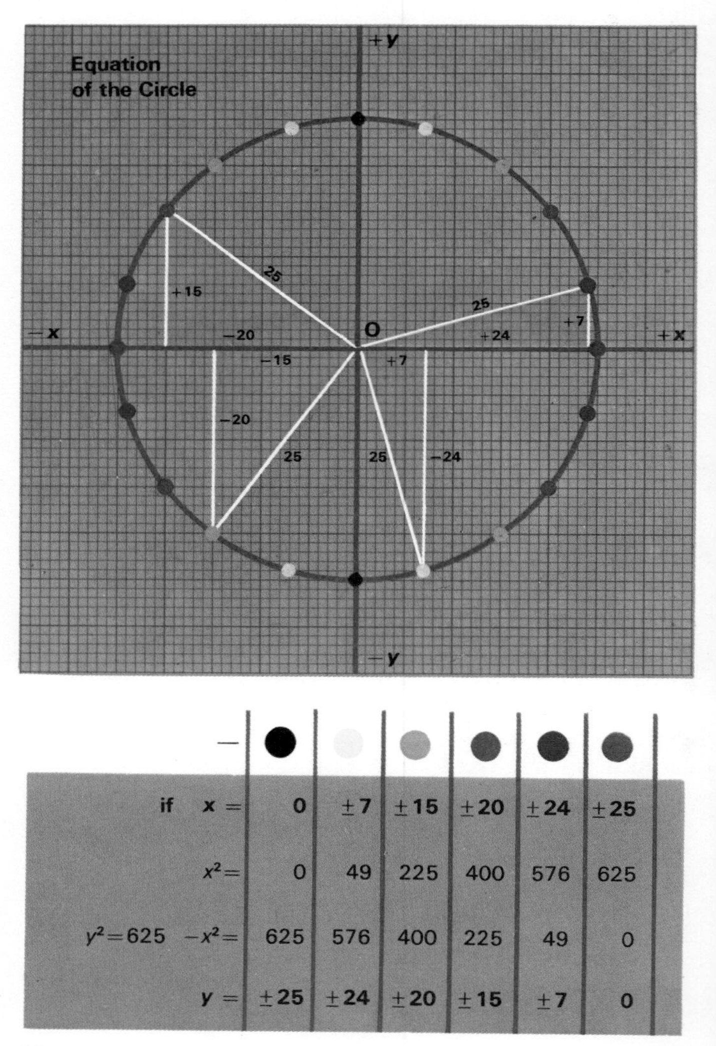

if $x =$	0	±7	±15	±20	±24	±25
$x^2 =$	0	49	225	400	576	625
$y^2 = 625 - x^2 =$	625	576	400	225	49	0
$y =$	±25	±24	±20	±15	±7	0

The ratio of the area of a circle to its radius squared (π) is approximately 3·1416. Close enough for most purposes.

An indication of the method by which π can be found is shown in the diagram below. A circle (we show here a quarter-circle) may be regarded as equivalent to a series of rectangles. In this quarter-circle of unit radius the radius is divided into five equal parts. The area of the exterior rectangles is greater than that of the quarter-circle by the mauve areas; the area of the interior rectangles is smaller by the dark-blue areas. The desired area is between the two.

AB, AC, AD and AE are all of unit length, and each division along AE is $\frac{1}{5}$. The area of the largest rectangle is thus $\frac{1}{5}$. Since AC $= 1$, the perpendicular from C to the base is $\sqrt{1^2 - (\frac{1}{5})^2} = \sqrt{5^2 - 1^2}$.

In the same way the perpendicular from D is $\sqrt{1^2 - (\frac{2}{5})^2} = \frac{\sqrt{5^2 - 2^2}}{5}$. Thus the total area of the exterior rectangles is

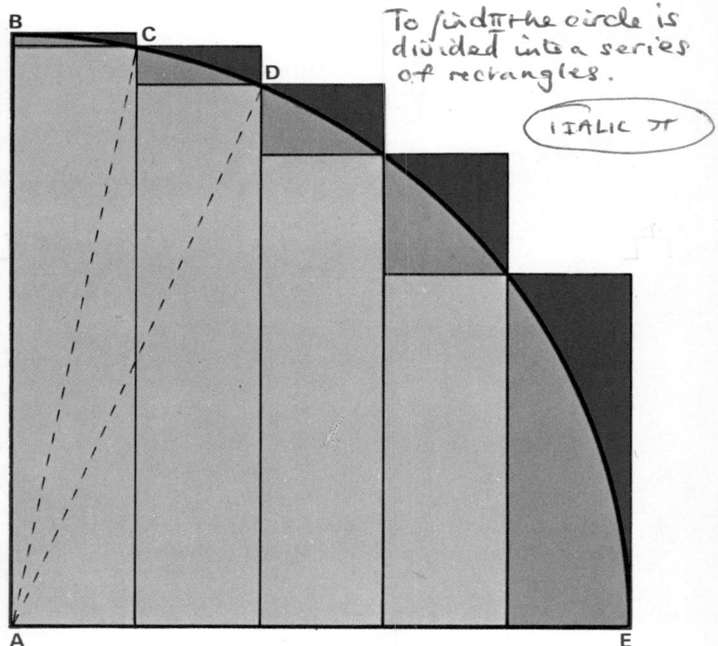

$\frac{1}{5^2}$ (5 + $\sqrt{24}$ + $\sqrt{21}$ + 4 + 3 = 1·72 approximately, and that of the inner rectangles $\frac{1}{5^2}$ ($\sqrt{24}+\sqrt{21}+4+3$) = ·66 approximately, which means that π (four times as great) lies between 3·44 and 2·64. That is not sufficiently accurate to be of value; but if the radius is divided into a large number of sections the difference between mauve and dark-blue portions will provide a good approximation.

More generally, the area of the exterior rectangles will be

$$\frac{1}{n^2}(n + \sqrt{n^2 - 1^2} + \sqrt{n^2 - 2^2} + \sqrt{n^2 - 3^2} + \ldots)$$

The 'imaginary' number

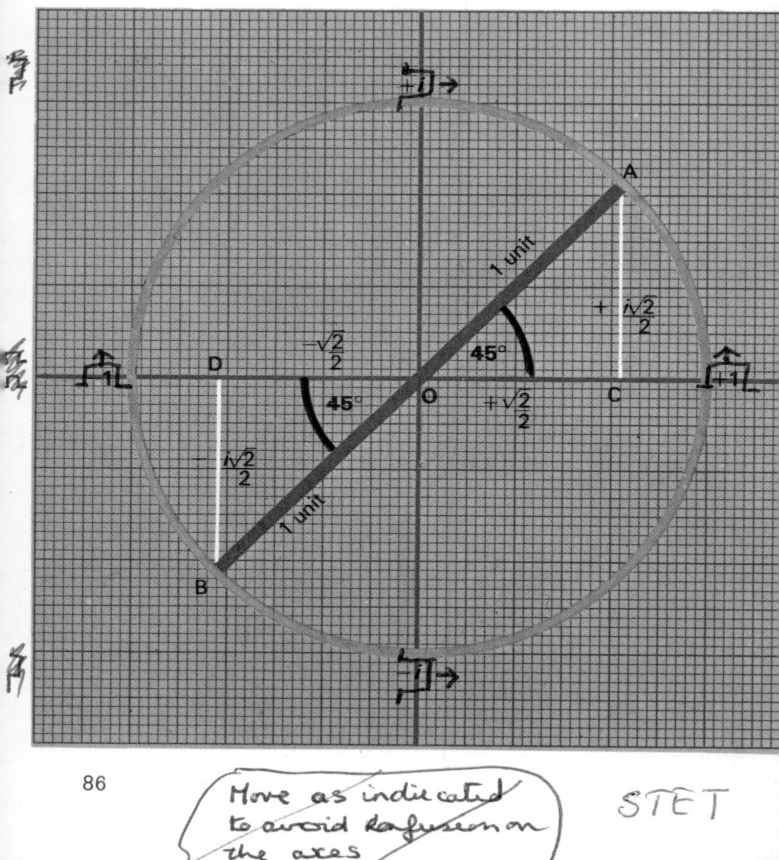

and that of the inner ones

$$\frac{1}{n^2}(\sqrt{n^2 - 1^2} + \sqrt{n^2 - 2^2} + \sqrt{n^2 - 3^2} + \dots).$$

These two series (giving the limits of $\frac{\pi}{4}$) can be summed by the integral calculus.

As has previously been explained, only points of which the co-ordinates are 'real' numbers can appear on a plane surface. But a graph can be used to *illustrate* complex and imaginary numbers. For this purpose we use the divisions on the x axis in the ordinary way to represent 'real' values and those on the y axis to represent imaginary numbers, that is to say, the vertical line three spaces to the right represents $+ 3$ and the horizontal line five spaces above the x axis represents $+ 5i$ (or $5\sqrt{-1}$).

Now, if we take a line of unit length having one end free and the other end anchored to the point O and let it revolve in an anti-clockwise direction it will trace out a circle, and we have already seen (see page 17) that when it has moved through 180° it will be multiplied by $- 1$, so that when it has moved through 90° its value will be $i\,(\sqrt{-1})$. What will be its value when it has moved through 45°? Clearly $\sqrt{i}$, or the square root of the square root of minus one. A graph can show what that is, see opposite.

AOC is an isosceles right-angled triangle, its hypotenuse being one unit. So each of the other two sides must be $+\frac{\sqrt{2}}{2}$ units, and the point A is therefore indicated by $+\frac{\sqrt{2}}{2} + \frac{i\sqrt{2}}{2}$. This is seen to be the case since:

$$\left\{\frac{\sqrt{2}}{2} + \frac{i\sqrt{2}}{2}\right\}^2 = \tfrac{1}{2} + \tfrac{1}{2}(i^2) + i = \tfrac{1}{2} - \tfrac{1}{2} + i = i.$$

In the triangle BOD the two legs of the triangle are $-\frac{2}{2}$ and $-\frac{i\sqrt{2}}{2}$, so that the point B is indicated by $-\frac{\sqrt{2}}{2} - \frac{i\sqrt{2}}{2}$. As before, we see that:

$$\left\{-\frac{\sqrt{2}}{2} - \frac{i\sqrt{2}}{2}\right\}^2 = \tfrac{1}{2} + \tfrac{1}{2}(i^2) + i = \tfrac{1}{2} - \tfrac{1}{2} + i = i.$$

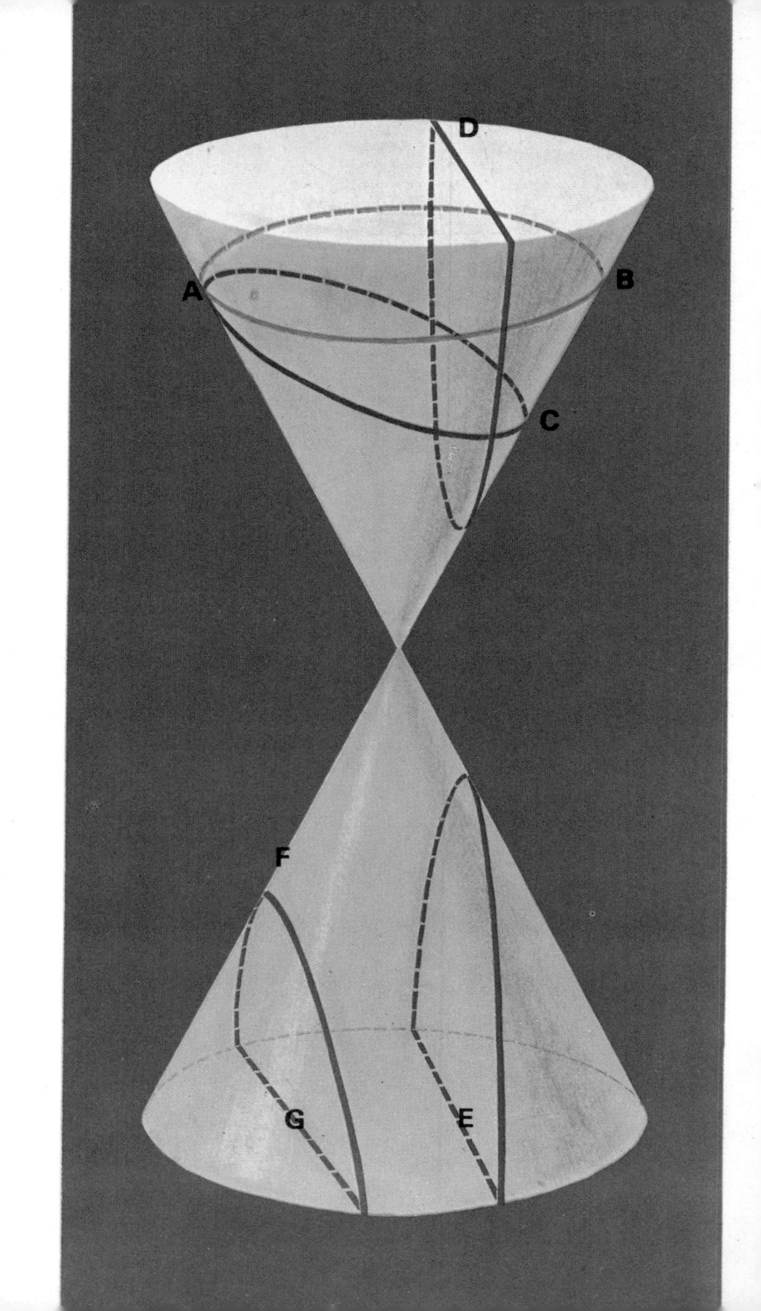
D
A
B
C
F
G
E

With reference to conic sections (previously mentioned), these are, as their name implies, derived from a cone, or rather from a double cone. If you imagine a cone inverted on the apex of another cone, you could cut this solid in a number of ways to produce various plane figures known as conic sections. In the diagram below the plane surface revealed by a cut from A to B would be a circle (as is also, of course, the base of the cone itself). An ellipse results from cutting from A to C; D to E produces a hyperbola; and F to G a parabola.

From two cones, one inverted upon the other, are derived the plane figures we call circle, ellipse, hyperbola and parabola.

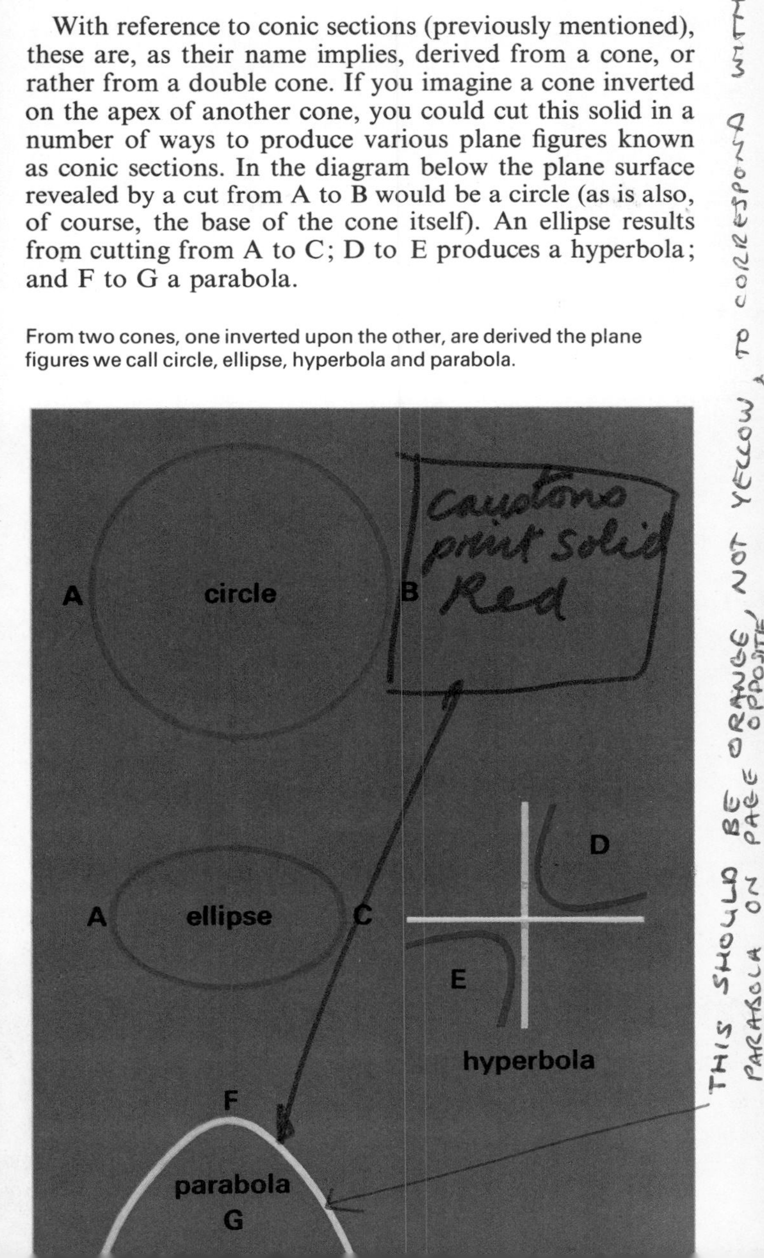

PROBABILITIES

'Probability is the very guide of life' – Joseph Butler

Towards the middle of the seventeenth century a French gambler, the Chevalier de Méré, who had been gaining a comfortable livelihood by laying the unwary even money that he would score at least one six in four throws of a single die, started to offer the same bet that he would score at least one double-six in 24 throws of two dice. Greatly to his discomfiture he found his profits beginning to dwindle, and he sought counsel from a friend, the great French mathematician and philosopher Pascal. The latter told de Méré that he was not the victim of any remarkable run of ill luck but of the immutable workings of probability; whereas one is likely to find a six in 3·8 throws of a single die, a double-six is likely to turn up only in 24·61 throws of two dice. The disgusted gambler then expressed the view that arithmetic was nothing but a swindle.

From this unimportant and even trivial event there developed, as the result of correspondence between Pascal and another great mathematician, Pierre de Fermat, the modern Theory of Probabilities, perhaps the field of mathematics that has made the greatest single contribution to the science of today.

For science has now learned humility. It has of course been known for centuries that not all physical laws apply invariably in all circumstances. There is a definite formula to express the expansion of a spring according to the weight attached to it – but only if the weight is not too light or too heavy. Statements that are true about physical objects cease to be true when those objects are enormously large or infinitesimally small. But although formulae were not all-embracing, this was, it was thought, because our mental powers are limited; so it followed that there was no reason in theory why a man of supreme ability should not be able to describe and even plot out the entire machinery of the universe.

Science until recent years stood solidly and, it seemed, immutably on the Law of Cause and Effect. If the

The Chevalier de Méré was a French gambler who wrongly calculated the probabilities at dice. When the philosopher Pascal showed him where he was going wrong the Chevalier declared that arithmetic was nothing but a swindle.

anticipated effect did not materialize, it was supposed, either there was some flaw in the experiment or else the cause was not identical with what we had supposed it to be; there could be no break in the logical chain. If, according to Laplace, one knew the whereabouts and the speeds and directions of every atom in the universe one

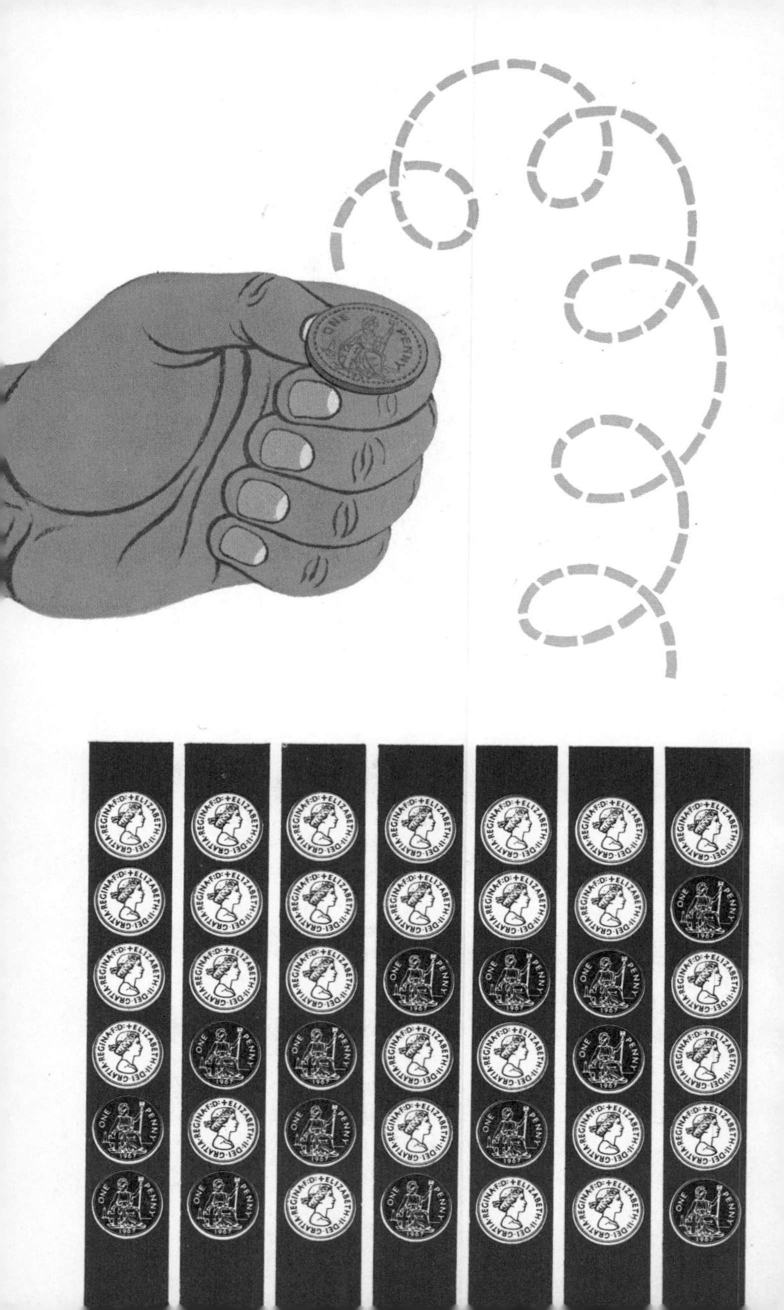
ONE PENNY
DEI·GRATIA·REGINA·F:D:+ELIZABETH·II
1967

could predict the future with certainty from now until eternity.

That is now known not to be the case. There is, for example, no way at all of determining the movement of individual electrons – although by taking a large number of them one may forecast the probable trend of the majority. This fact was embodied in 1927 by the German physicist Heisenberg in what he called the Principle of Uncertainty (a principle known to artists if not to scientists throughout the ages). And the one branch of mathematics specifically concerned with the principle of uncertainty is the Theory of Probabilities.

This theory (not, if you please, the 'laws of chance' – if chance had laws it wouldn't be chance) is sometimes called the Theory of Large Numbers, perhaps because it doesn't work out at all *except* with large numbers (we shall come back to this point when we talk about Statistics).

Everybody knows, for example, that if you toss a penny it is as likely to come down heads as tails, so that if you toss a penny *a great many times* you are likely to get *about*

If you toss a penny six times and two of the tosses turn out to be tails, there are 15 ways in which you may have got this result just as there are 15 ways of throwing four tails and two heads.

as many heads as tails. But it must be a great many times, and you must not expect the result to agree exactly with your forecast. The larger the number of tosses the greater the measure of agreement there is likely to be. Obviously, for example, if there is an odd number of tosses the numbers of heads and tails cannot be exactly equal. But supposing you toss your penny six times, are you likely to get just three heads and three tails? The answer is no. Without going into any complicated mathematics, it is quite easy to list the possibilities (all of which are equally likely). There are two possible results for the first toss (head or tail), the same for the second, and so on. So the total number of possible results is 2^6, or 64. Two of these will be all heads or all tails; if only one tail turns up it may be the result of any one of the six tosses, so there are six ways of getting only one tail (and, of course, six ways of getting only one head). Where two of the tosses turn out to be tails, the possible sequences are as illustrated on the preceding page.

Thus there are 15 ways of throwing four heads and two tails, and, of course, 15 ways of throwing two heads and four tails. So we have accounted for $2 + 12 + 30 = 44$ out of our possible 64 sequences, leaving only 20 ways of throwing three heads and three tails. So, although you are likely to get more results of three heads, three tails than of four heads, two tails, the three-three result is less likely than getting *one* of the *two* combinations four heads and two tails and two heads and four tails.

If you toss a coin a thousand times you are not very likely to get *exactly* 500 heads and 500 tails, but it is extremely improbable that the result will vary much.

Your probability of tossing exactly one, two, three ... heads out of so many coins is illustrated by a contrivance of Pascal. Known as Pascal's Triangle, it is shown opposite.

Pascal's Triangle is quite easily formed. For the first line one writes the figure 1 twice. For the second and all succeeding lines one starts (and finishes) with unity and for each of the other figures writes the sum of the two figures above it; thus 1 1 is followed by $1\ (1 + 1)\ 1 = 1\ 2\ 1$, and this in turn is followed by $1\ (1 + 2)\ (2 + 1)\ 1 = 1\ 3\ 3\ 1$, and so on.

Pascal's triangle

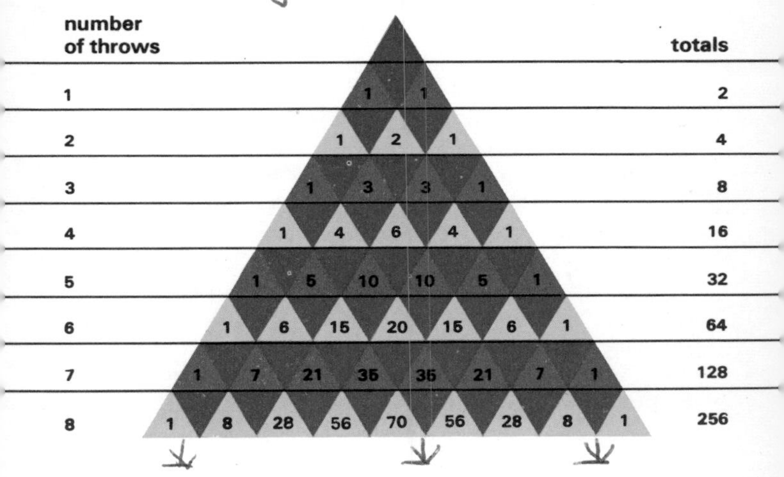

This provides a very simple method of estimating your probability of throwing x heads (or tails) out of n tosses of one coin: you merely read across the nth line of the triangle. If there are to be, say, four tosses, then your probabilities are found by reading across line 4: there is one chance out of 16 that you will throw four heads; there are 4 chances out of 16 that you will throw three heads (and one tail); 6 chances of throwing two heads (and two tails); 4 of getting one head (and three tails); and 1 of throwing all tails.

If you toss eight times, read across line 8: your probability of throwing all heads or all tails is $\frac{1}{256} + \frac{1}{256} = \frac{1}{128}$; of throwing 7 heads and 1 tail or 1 head and 7 tails is $\frac{8}{256} + \frac{8}{256} = \frac{1}{16}$; of throwing 6 heads and 2 tails or 2 heads and 6 tails is $\frac{28}{256} + \frac{28}{256} = \frac{7}{32}$; of throwing 5 heads and 3 tails or 3 heads and 5 tails is $\frac{56}{256} + \frac{56}{256} = \frac{7}{16}$; and of throwing 4 heads and 4 tails is $\frac{70}{256} = \frac{35}{128}$.

Obviously you *must* throw one of these combinations, so that the fractions denoting the probabilities should add up to unity (certainty). It will be seen that they do:

$$\frac{1}{128} + \frac{1}{16} + \frac{7}{32} + \frac{7}{16} + \frac{35}{128} = 1.$$

You will note that in all cases (except where you confine yourself to only two throws of a coin) the probability of getting exactly the same number of heads and tails is less than $\frac{1}{2}$. This has consequences (particularly in statistical theory) of far greater importance than in gambling.

Pascal's Triangle is interesting for other reasons. It will be seen that, reading diagonally downwards from the apex, the first line is a succession of ones; the second is the series of natural numbers, and in this and all subsequent lines each term is the sum of all earlier numbers in the preceding line. They are, in fact, 1, 1, 1, 1 ...; 1, (1 + 1), (1 + 1 + 1), (1 + 1 + 1 + 1) ... = 1, 2, 3, 4 ...; 1, (1 + 2), (1 + 2 + 3), (1 + 2 + 3 + 4) ... = 1, 3, 6, 10 ...; 1, (1 + 3), (1 + 3 + 6), (1 + 3 + 6 + 10) ... = 1, 4, 10, 20 ..., and so on. Furthermore, these are found to be the coefficients of x in the expansion of the binomial $(x + 1)^n$, which is $x^n + \frac{nx^{n-1}}{1.} + \frac{n(n-1)x^{n-2}}{1.2.} + \frac{n(n-1)(n-2)x^{n-3}}{1.2.3} + \ldots$ Taking the case where $n = 4$ (line 4 in the Triangle), this expansion becomes $x^4 + 4x^3 + 6x^2 + 4x + 1$, the coefficients thus being 1, 4, 6, 4 and 1.

Probabilities are widely, and not very successfully, applied to gambling; not very successfully because not very scientifically. Take as an example the game of roulette, for which a number of 'systems' are used. It is noteworthy that while suicides by ruined gamblers are not uncommon casinos continue to flourish. The reason is that the proprietors of the casinos apply the theory of probabilities correctly and the punters don't.

A standard European roulette board employing one zero. The presence of the zero is sufficient to shade the odds in favour of the house by 37 to 36 on any single-number bet.

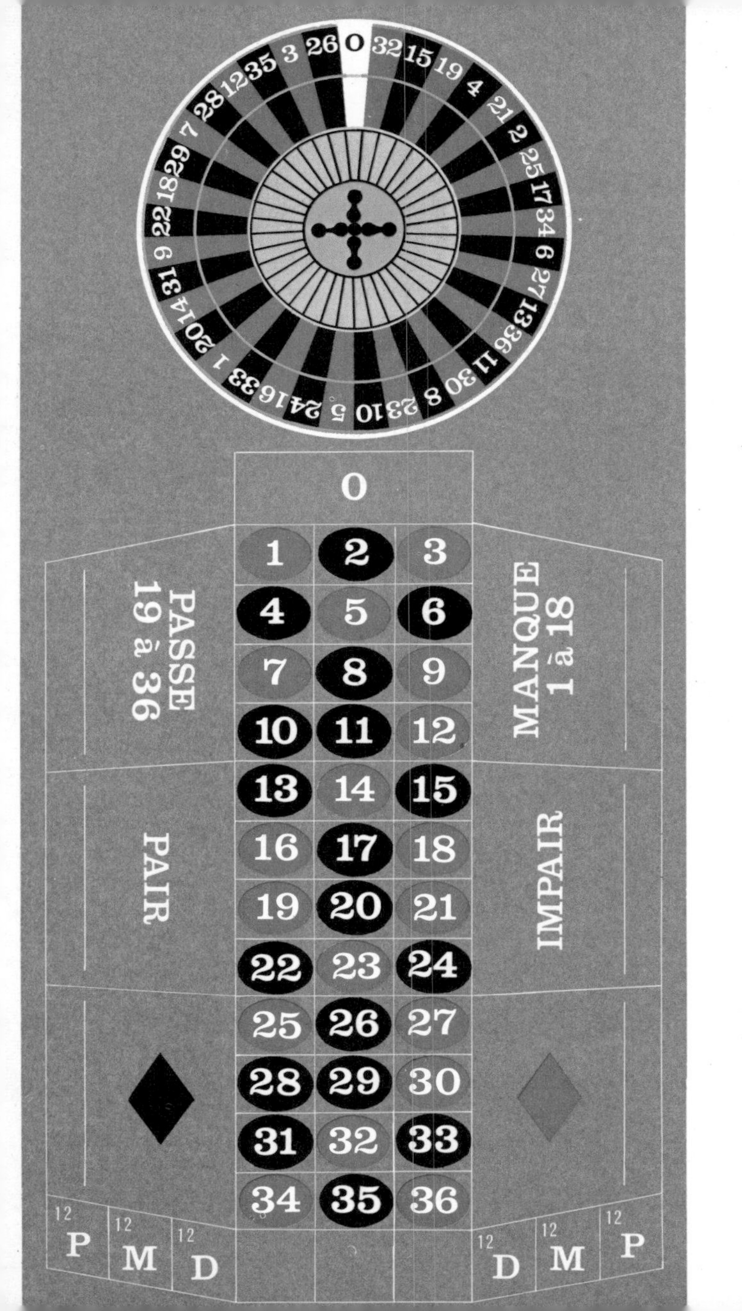
0
1 2 3
4 5 6
7 8 9
10 11 12
13 14 15
16 17 18
19 20 21
22 23 24
25 26 27
28 29 30
31 32 33
34 35 36
PASSE 19 à 36
MANQUE 1 à 18
PAIR
IMPAIR
12 P
12 M
12 D
12 D
12 M
12 P

The roulette board used in Monte Carlo and in the south of France is numbered from 1 to 36, 18 of the numbers being red and 18 black, and there is also a zero.* There are many ways of betting at roulette, but we will confine ourselves to the so-called even chances: black or red, odd or even, high or low. Actually these are not quite even chances; if zero turns up all bets other than those on zero are lost with the exception of the 'even' ones – they remain on the board until after the next spin, when, according to the result, the punter's stake is lost or may be taken back. Thus, all possible results being equally likely, you may normally expect any number (including zero) to turn up on an average twice in 74 spins, giving you 36 wins, 37 losses and 1 stake returned, so that the odds against you are 37 to 36. A very small balance indeed, but enough to ensure that in the long run the bank will win. That shade of odds covers all the overheads of the casino plus a handsome profit for the shareholders.

That is how the theory of probabilities works. To see how it doesn't work, study the activities of 'system' fans.

Many of them fall into the obvious error of supposing that because red, for example, has turned up half a dozen times in succession it is now 'black's turn', and push their plaques into the black section. But the past has no influence on the future. Black is no more likely to turn up even if there have been twenty successive reds.

Another supposed method of beating the bank is what is known as the Martingale. It has an engaging appearance of infallibility. All you do is to back, say red, consistently and double your stakes each time you lose, reverting after a win to your original stake. Thus every red shows you a profit of one plaque over and above whatever you may have lost on that series.

All very easy, and at first glance convincing. Indeed, the Martingale would be an infallible method, granted that the bank will without limit cover any bet you like to stake and

* Roulette boards in some other parts of the world have two or even three zeros. Rather than play on these, throw your money into the sea. It is an equally amusing and effective way of getting rid of it.

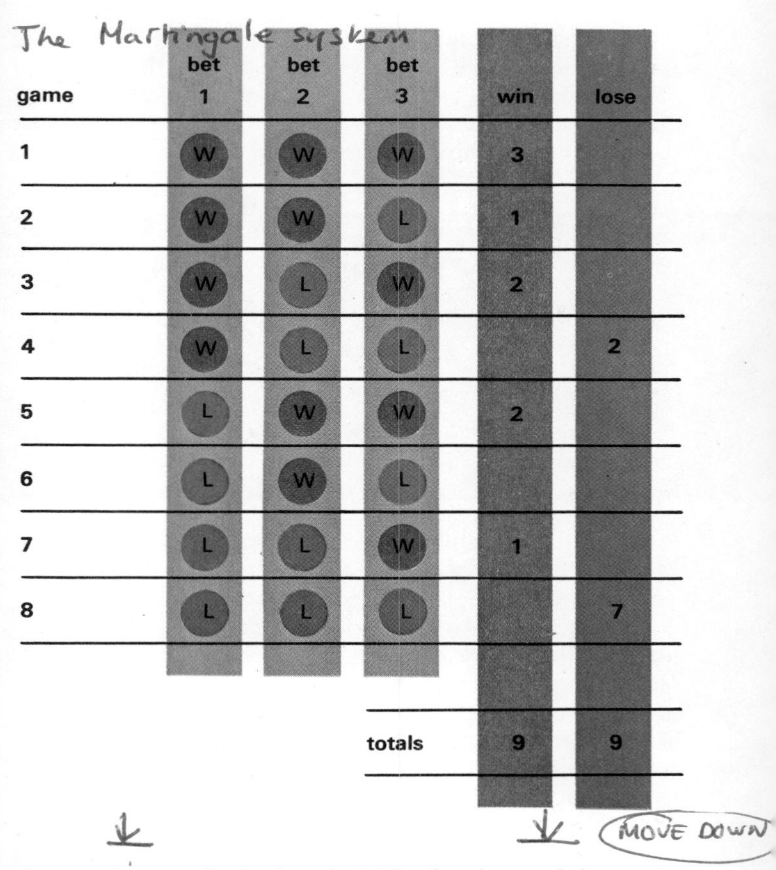

game	bet 1	bet 2	bet 3	win	lose
1	W	W	W	3	
2	W	W	L	1	
3	W	L	W	2	
4	W	L	L		2
5	L	W	W	2	
6	L	W	L		
7	L	L	W	1	
8	L	L	L		7
			totals	9	9

that you have unlimited capital. The first is certainly not the case; the second is, I suspect, unlikely.

Let us see how the Martingale fares over three bets. These may result in any of eight possible ways (excluding the possibility of zero turning up and assuming that your chances are in fact even, as our illustration shows).

In game no. 1 you win three units, in no. 2 one unit, in no. 3 two, in 4 you lose two, in 5 you win two, in 6 you neither win nor lose, in 7 you win one, and in 8 you lose seven. So total winnings are 3 + 1 + 2 + 2 + 1 = 9, and your losses are 2 + 7 = 9; you are neither better nor worse off than if you had bet a level stake throughout.

In five bets, again ignoring the possibility of zero turning up, you may expect out of 32 games (two of which will be inconclusive) to show 4 wins of one stake, 7 of two, 7 of

If there are ten horses in a race . . .

three, 4 of four, and 1 of five, a total of 60, and to lose 2 of one stake, 1 of two, 1 of five, 1 of six, 1 of fourteen and 1 of thirty-one, again a total of 60.

Comparing Martingale with ordinary betting, you are likely to win more often than you will lose, but to lose larger amounts; and if your run of bad luck comes early in the proceedings you may well find your capital so diminished that you cannot wait for fortune to take a more favourable turn. 'Doubling up' runs into very big money very quickly. If, backing red with an initial stake of one pound, you experience an unbroken run of ten blacks (by no means an exceptional occurrence) then for the eleventh bet your Martingale will require a stake of £1,024 if you are to avoid a heavy loss, and therefore an initial capital of £2,047. Better go to the cinema instead!

All this is, of course, not to say that an intelligent study of probabilities will not help you in games of chance; it will, within limits. There are tables to tell you the probable distribution of the trumps at bridge, and in poker the likelihood of filling a straight or a flush; these the player will disregard at his peril. But both bridge and poker are games of skill as well as of chance; in particular the poker

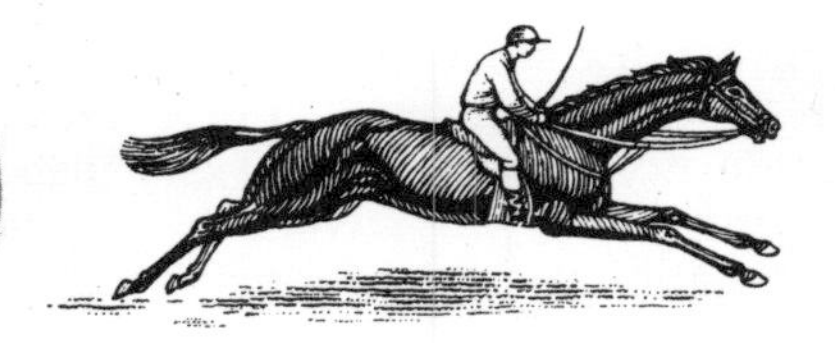

. . . to forecast
the first three horses
with certainty
would require

$$\frac{10 \times 9 \times 8}{1 \times 2 \times 3}$$

= 120 bets

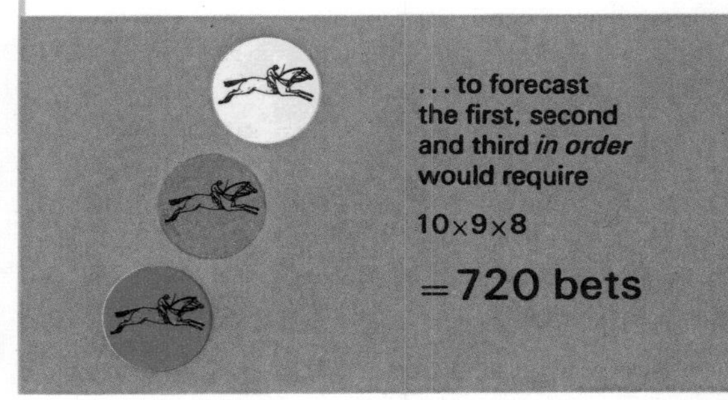

player who works purely on mathematical principles is a certain loser. He must rely on psychological probabilities, or 'hunches', to decide if an opponent is bluffing.

The theory of probabilities is largely based on the mathematical technique of permutations and combinations. A combination is the number of ways of selecting x things out of n things (this is the method you use in filling in what the pools promoters, for reasons best known to themselves, call a 'perm'). A permutation is the number of ways of selecting x things *in a definite succession* out of

n things, as when you attempt to nominate the three placed horses in a race in their order of precedence. It would be a combination if you selected the three placed horses without forecasting the order in which they would finish.

A permutation is the product of the x numbers n, $n-1, n-2, n-3$, and so on; a combination is the same product divided by 1.2.3.4...x times. Thus, to forecast the winner, the second and the third horse in a field of ten runners with certainty would require $10.9.8 = 720$ bets, whereas in a football Treble Chance the combination (*not* the 'perm') needed to pick the required 8 matches out of 11 requires $\frac{11.10.9}{1.2.3} = 165$ lines. Note that we have here used only three terms instead of the eight you might have expected, the reason being that to *choose* 8 teams out of 11 is the same thing as to *reject* 3 teams out of 11, and $\frac{11.10.9}{1.2.3}$ is actually the same thing as $\frac{11.10.9.8.7.6.5.4}{1.2.3.4.5.6.7.8}$; the last five numbers in the numerator cancel out the last five in the denominator.

It may be thought that in this section too much attention has been paid to the application of the theory of probabilities to games of chance. But, apart from the fact that the theory is there most clearly illustrated, most of our activities in life are perforce in part gambles. When you insure against burglary you are in fact betting the insurance company that your house will be burgled and the company offers you odds that it will not. When you take out a life assurance policy you are not, of course, betting that you will die – that is one of the few things of which you can be certain – but that you will die sooner than the company supposes. When you take out an annuity, on the other hand, the company is betting on your early demise. The company, with the advice of actuaries, is fixing the odds; that is why they make the profits.

Probability alone will not tell you whether a bet, an investment or any other kind of business deal is likely to be profitable to *you*. In the case of burglary insurance, for example, you will probably pay premiums for many years and never make a claim, so you will be out of pocket; but

if you *should* be burgled you will have avoided what might have been a heavy loss. Apart from mathematical probabilities, how well are you able to stand this problematical loss? If you were to offer to lay me two to one in shillings against my tossing a head in a single throw of a coin I should be most happy to accept the bet. But if a mad millionaire offers to bet me £10,000 to £5,000 on the same chance I shall decline with thanks. While £10,000 would be an extremely welcome windfall, it wouldn't transform my life into a dream of bliss; whereas the loss of £5,000 would bankrupt and ruin me. It is a good bet mathematically, but a very bad bet for a man in my particular circumstances.

So regard probabilities as a guide – perhaps the best we have in mundane affairs – but not an infallible signpost to an inevitable future. That way lies disappointment; and you may find yourself agreeing with the Chevalier de Méré that arithmetic is 'nothing but a swindle'.

HOW THINGS GROW

'I 'spect I growed' – Topsy in Uncle Tom's Cabin, *Harriet Beecher Stowe*

If you were able to invest your money at 100 per cent compound interest per year it would be highly advantageous. However fantastic the idea may appear it will lead us to matters that are severely practical.

One pound invested at 100 per cent for a year would double itself (just as it would if it were invested at simple interest). But suppose dividends are declared half-yearly. Then compound interest *would* make a difference: you would be due to receive (and re-invest) ten shillings at the end of six months, so for the next half-year interest at the rate of 50 per cent would be accruing on the enhanced capital of £1 10s. As a result by the end of the year your principal of one pound would have increased to £2 5s, see the table opposite.

Now suppose the company declares dividends three times a year. You will be doing better still; you will get $33\frac{1}{3}$ per cent interest on your pound for four months, another $33\frac{1}{3}$ on £1 3s 4d, and a third $33\frac{1}{3}$ per cent interest on the sum to which your capital has increased after eight months. There is a simple formula to calculate the increase in capital invested at 100 per cent compound interest. Taking P as the principal, $\frac{1}{n}$ as the fraction of it accruing during a certain period and n as the number of such periods, the increased capital will amount to $P\left\{1+\frac{1}{n}\right\}^n$. Since we are working on a capital of £1, $P = 1$ and can be ignored. We can check this from the first two examples:

when $n = 1$, $\left\{1+\frac{1}{n}\right\}^n = 2$ (capital increased to £2);

when $n = 2$, $\left\{1+\frac{1}{n}\right\}^n = (1+\frac{1}{2})^2 = 2\frac{1}{4}$, or £2 5s.

If we regard interest as accruing *constantly*, then there is an unlimited number of periods to be considered and a

Compound Interest

If dividends are declared twice a year

Principal	**£1**	**0**	**0**

After 6 months
add ½ year's interest
£1 at 50%

	10	**0**
£1	**10**	**0**

After 12 months
add ½ year's interest
£1 10 at 50%

15	**0**

Total at end of
12 months

If dividends are declared three times a year

Principal	**£1**	**0**	**0**

After 4 months
add ⅓ year's interest
£1 at 33⅓%

	6	**8**
£1	**6**	**8**

After 8 months
add ⅓ year's interest
£1 6 8 at 33⅓%

	8	**10**
£1	**15**	**6**

After 12 months
add ⅓ year's interest
£1 15 6 at 33⅓%

11	**10**

Total at end of
12 months

£1 @ 100% =

£2.5.0 £2.7.4

very much diminished rate, so that our formula seems to give a result approaching $\left\{1 + \frac{1}{\infty}\right\}^{\infty}$ – which is of course quite meaningless.

Perhaps the easiest way of getting over the difficulty is to see what actually *does* happen when the number of periods increases and the rate of interest diminishes accordingly, see the table opposite.

It seems pretty clear that this series is not going to reach infinitely large proportions however far on we carry it; although it increases, the amounts by which it grows become progressively smaller. We expect to find it a convergent series that will never exceed a certain figure. That figure is in fact 2·71828...

This number is known to mathematicians as e, the sum of the exponential series, and it is important in many kinds of calculation. The series is:*

$$1 + \frac{1}{1!} + \frac{1}{2!} + \frac{1}{3!} + \frac{1}{4!} + \ldots$$

A simple (but to my mind highly unsatisfactory) way of finding the value of $\left\{1 + \frac{1}{n}\right\}^{n}$ when n is infinitely great is given in some text-books. It is by means of the binomial theorem (see Appendix II).

We know that $(1 + a)^n$ when n is finite is expanded to make

$$1 + na + \frac{n(n-1)}{2!}a^2 + \frac{n(n-1)(n-2)}{3!}a^3 + \ldots$$

so that substituting $\frac{1}{n}$ for a, we have then the following equation

$$\left\{1 + \frac{1}{n}\right\}^{n} = 1 + \frac{n}{n} + \frac{n(n-1)}{n^2.2!} + \frac{n(n-1)(n-2)}{n^3.3!} + \ldots$$

But if n is infinite then $n - 1$, $n - 2$, etc. must also be infinite, so that every term in the numerator containing n

* An exclamation mark after a number signifies that it is a factorial; that is to say it represents that number multiplied by every number smaller than itself. For example, $4! = 4.3.2.1 = 24$.

number of interest periods per year	capital increases to	to 5 places of decimals
1	$\{1 + \frac{1}{1}\}^1$	2
2	$\{1 + \frac{1}{2}\}^2$	2·25
3	$\{1 + \frac{1}{3}\}^3$	2·37037
4	$\{1 + \frac{1}{4}\}^4$	2·44141
5	$\{1 + \frac{1}{5}\}^5$	2·48832
6	$\{1 + \frac{1}{6}\}^6$	2·52163
7	$\{1 + \frac{1}{7}\}^7$	2·54650
8	$\{1 + \frac{1}{8}\}^8$	2·56578
9	$\{1 + \frac{1}{9}\}^9$	2·58117
10	$\{1 + \frac{1}{10}\}^{10}$	2·59374

As the number of interest periods per year increases the rate of interest diminishes accordingly, reaching an eventual limit at 2·71828. . . . This limit is known to mathematicians as *e*, the sum of the exponential series.

can be 'cancelled out' with an *n* in the denominator, thus leaving us with the result:

$$\left\{1 + \frac{1}{\infty}\right\}^{\infty} = 1 + 1 + \frac{1}{2!} + \frac{1}{3!} + \ldots$$

The result is correct; but the means by which it is reached are not likely to commend themselves to the mathematician who seeks a rigorous demonstration. For example, the third term in the series $\left\{\frac{n(n-1)}{n.2!}\right\}$ is equated with $\frac{1}{2!}$ by contending that $\frac{\infty \times \infty}{\infty \times \infty \times 2!} = \frac{1}{2!}$, thus 'cancelling out' infinities. This is dangerously similar to the 'schoolboy howler' on page 60, that is, the proof that 1 = 2. It illustrates the fact that arriving at a correct conclusion does not necessarily validate the reasoning that precedes it.

A better course is to sum the series for progressive numbers of terms (calculated to four places of decimals), as shown opposite.

Now this series we *know* to be convergent, because the tenth term will require an addition of $\frac{1}{10!}$, the eleventh one of $\frac{1}{11!}$, and so on, each successive increase being less than one-tenth of its predecessor, and we remember (referring back to Geometrical Progressions) that $x + \frac{x}{10} + \frac{x}{100} + \ldots$ can never exceed $x(1{\cdot}\dot{1})$.

The exponential series is vital to the understanding of *growth*, because although investments do not increase constantly in this way, living things do, and growth (of population, for example) can often be expressed as some function of the exponential number.

The number e is unique in another way. We can express any power of it (that is to say, any number of e's multiplied together) by a very simple formula. If x is the required power, then

$$e^x = 1 + x + \frac{x^2}{2!} + \frac{x^3}{3!} + \frac{x^4}{4!} + \ldots$$

This is true for *any* value of x, positive or negative, integral or fractional. Let us try a couple of examples. When $x = 2$,

$$e^2 = 1 + 2 + \frac{4}{2!} + \frac{8}{3!} + \frac{16}{4!} + \frac{32}{5!} + \frac{64}{6!} + \ldots$$

$$= 1 + 2 + 2 + 1{\cdot}\dot{3} + {\cdot}\dot{6} + {\cdot}2\dot{6} + {\cdot}0\dot{8} + \ldots = 7{\cdot}385$$

to seven terms, and $(2{\cdot}718)^2 = 7{\cdot}3875\ldots$

number of terms in the series	sum	add to find next term
1	1	1
2	2	$\frac{1}{2!} = \cdot 5$
3	2·5	$\frac{1}{3!} = \cdot 1667$
4	2·6667	$\frac{1}{4!} = \cdot 0417$
5	2·7084	$\frac{1}{5!} = \cdot 0083$
6	2·7167	$\frac{1}{6!} = \cdot 0014$
7	2·7181	$\frac{1}{7!} = \cdot 0002$
8	2·7183	etc.

It is obvious that if we had continued our series a little further we should have come close to the desired result.

Equally with a negative index:

$$\frac{1}{e} = e^{-1} = 1 + 1(-1) + \frac{1(-1)^2}{2!} + \frac{1(1-1)^3}{3!} + \frac{1(-1)^4}{4!} + \frac{1(-1)^5}{5!} + \frac{1(-1)^6}{6!} + \ldots$$

$$= 1 - 1 + \tfrac{1}{2} - \tfrac{1}{6} + \tfrac{1}{24} - \tfrac{1}{120} + \tfrac{1}{720} - \ldots$$

$$= \text{approximately, } \tfrac{53}{144} = \cdot 368,$$

and $$\frac{1}{2 \cdot 718} = \cdot 368.$$

It was this peculiarity of the exponential series that led Napier to adopt e as the base of his 'natural' logarithms instead of ten as used in common logarithms.

WHAT MACHINES CAN, AND CANNOT, DO

'It is the Age of Machinery, in every outward and inward sense of that word' – Thomas Carlyle

The Machine that can Think is a bogey-man that has haunted us over the centuries. It appears constantly in myth and legend; Karel Čapek's play 'Rossum's Universal Robots' is one example in recent literature. Then the Thinking Machine found its natural habitat in the pages of the horror comics, whence it has in comparatively recent days emerged to be the subject of apparently grave discussion in the columns of supposedly serious newspapers. This is a pity. The Thinking Machine should go back to the horror comics.

Anybody who talks about a Thinking Machine either doesn't know what thinking is or else doesn't know what a machine is. A machine is a tool operated by a man; it will, if it is in sound working order, carry out its instructions efficiently – as efficiently as its operator directs it. Machines, we are told, don't 'make mistakes'. Of course they don't; they don't make anything. If any errors have crept into this book (a not impossible hypothesis) there will be little use in my telling indignant readers that my typewriter has let me down. Typewriters don't make spelling mistakes; but typists sometimes do.

It would be ridiculous to deny that machinery, and particularly electronic machinery, has in the past few years made fantastic progress; but there has not been and never will be a break-through into conscious and volitional activity. An electronic computer is, when you get down to fundamentals, only an improved abacus.

A great deal of the current misunderstanding about machinery arises from the misuse of metaphors. We say that machines have 'memories' and that they can make 'decisions'. What we mean in fact is that they can store information (as can a filing cabinet) and that they respond

Calculation through the ages, from the abacus to the computer; in between, a seventeenth-century version of the slide-rule.

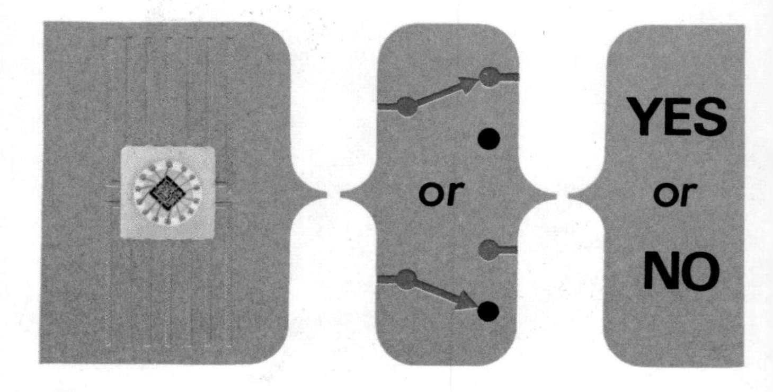

A micro-miniature circuit used in modern computers. The circuit performs given logic functions and will supply 'yes' and 'no' answers according to information supplied to the computer.

to stimuli. The thermostat is a useful and valuable invention; but nobody would venture to suggest that it 'decides' to switch off the heat when a certain temperature is reached.

Electronic machinery works at lightning speed and can reply in a matter of seconds to questions that might engage a team of mathematicians for years; but it can deal only with a restricted range of questions, and it has to be 'programmed'. It never starts from scratch; information must be fed in before it can be given out. Also, since the machine works on impulses, the only answers it can give to any question are 'yes' and 'no' corresponding to the positive and negative states of 'on' and 'off'. As anyone who has listened to the radio game 'Twenty Questions' will be aware, this is not the most satisfactory way of eliciting information.

Particularly in the field of mathematics, information can be obtained from a machine only if the question is first framed in mathematical language. No machine can solve a 'problem'. Let us consider for a moment what we do when we deal with problems – and first take one so simple that it hardly rates the appellation at all. John has six apples, Henry has two. How many apples must John give to Henry so that each of them shall have the same number?

First, how do *we* deal with the matter? If we decide to give this triviality 'the full treatment' we say: John must give Henry x apples, with the result that he will then have $6 - x$ whereas Henry will have $2 + x$. Therefore $6 - x = 2 + x$, so that $4 = 2x$ and $x = 2$. So John hands over two apples.

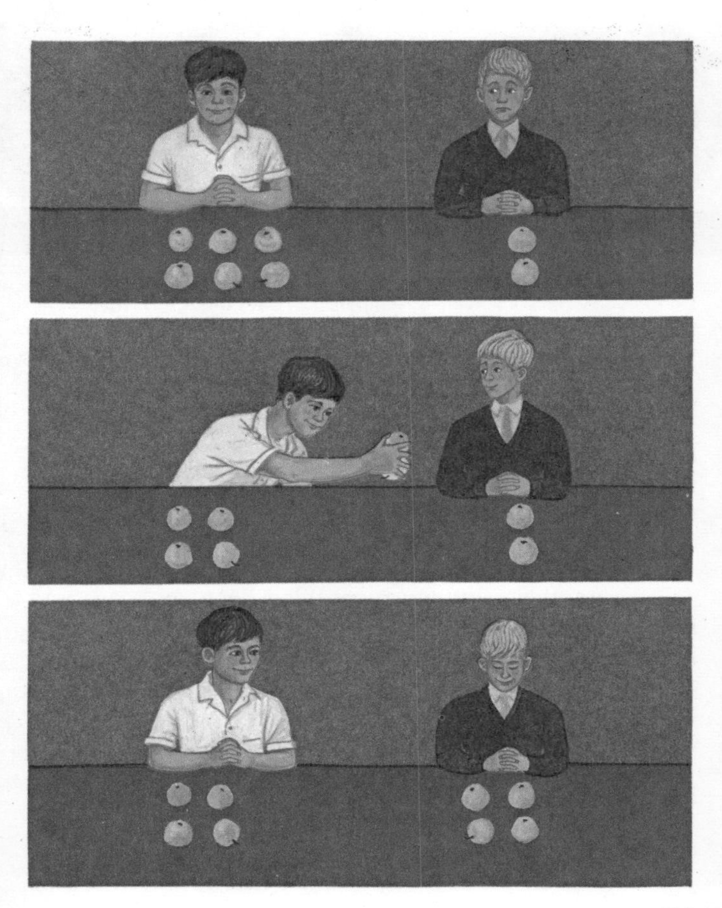

Contrary to what some wary bystanders may have thought as they watched its erratic progress, the early motor-car was controlled exclusively by the man behind the wheel.

There are actually three steps here: first we translate the situation into algebraic symbols, next we solve the equation, and finally we translate the answer back again into terms of boys and apples. The second step you may confidently entrust to a machine; but you will find it wholly incapable of dealing with either of the others.

Take another very simple problem. My son's age and mine together total 65 years and they differ by 25 years. How old are we? How do you do this? You say: let my age in years be x and my son's y. Then $x + y = 65$ and $x - y = 25$, so that $x = 45$ and $y = 20$; thus I am 45 years old and my son is 20 years old.

Correct? Not quite: there is a logical step omitted. We have said, without a second's thought, that $x - y = 25$; but why so? Why not $y - x = 25$, making my son's age 45 and mine 20? This you will reject indignantly because you know (being a thinking animal) that a father must be

Machines have never taken us over and there is no reason to suppose that they ever will–despite the modern-day superstition that computers are Thinking Machines.

older than his son. But a machine 'knows' nothing outside the facts with which it has been 'programmed'. The machine not only *might* but *should* give you an alternative solution to the question.

It may well be that electronics will radically transform our lives. So did the internal combustion engine; but we never fell into the trap of regarding that machine anthropomorphically. The 'brain' of a modern computer (one can hardly escape these false analogies) is a vastly complicated system that may occupy a space as large as a room. You have in your skull an infinitely more complicated system in a double-handful of matter – a system not only more complicated but capable of feats that are not merely outside the range of a computer but wildly beyond anything it might conceivably be supposed to be able to achieve in the far-distant future.

There is only one kind of Thinking Machine. It is you.

HOW STATISTICS CAN FOOL YOU

'There are three kinds of lies: lies, damned lies, and statistics' – *Mark Twain*

You are likely to come across statistics from two sources: learned publications and the advertisement columns of your daily newspaper. They differ considerably in many respects, but they have this much in common: neither of them (*pace* Mark Twain) is likely to be composed of flat lies. That is to say, if a scientific commission tells you that a thousand mice contaminated with some disgusting compound died, then – whatever may be the value of the information – they almost certainly did. Equally when you read in an advertisement that eight film-stars out of ten clean their teeth with SHEENO you may be reasonably confident that there are in fact ten film-stars of whom eight use SHEENO. Should you have assumed that eight out of *every* ten film-stars (or 80 per cent of all of them) are SHEENO addicts, why, that is your fault. The advertisement didn't say so.

Statistics can be misleading in a number of ways: first and foremost, they are utterly valueless unless they cover a large and random selection of cases. Suppose, for instance, you are told that during one year in a certain hospital one hundred per cent of unvaccinated patients died of smallpox whereas among vaccinated patients there was not a single fatality, you will probably draw some very far-reaching conclusions. But let us suppose you now learn that this hospital during the year under review admitted precisely two cases of smallpox. That will surely give you pause. A hundred per cent sounds very large indeed; but when considering only one patient of each category it is just one more than nought per cent: the difference is statistically speaking not significant.

Of no less importance is the 'randomizing' of the cases considered. Suppose out of 40 patients suffering from a particular disease 20 are given the latest 'wonder drug' while their fellows receive only the orthodox treatment, and that cures among those taking the new preparation

(Opposite) dazzling the public with statistics

TAKE 2 LIN
ON

8 out of 10
use SH

'Over a certain period of time statistical evidence has indicated that cigarette smoking may cause lung cancer. Over the same period of time there has been an enormous increase in the use of television aerials. Therefore television causes lung cancer.' Obviously this is not true, but statistics are sometimes misused in this way, linking data to present a new 'fact' where no causal connection exists.

outnumber cures among the others by three to one – is this significant or not? First we must know a little more. What if the 'wonder drug' patients were all under twenty years of age while their fellow-patients were septuagenarians? Will we not then have second thoughts – that is, of course, if we are ever made aware of this age differential?

I have a veneration bordering on the superstitious for all doctors; they know so very much more than I do about medicine, about anatomy, about hygiene. But do they,

I sometimes wonder, know so very much more than I do about mathematics? Because statistics is a branch – and by no means the easiest branch – of mathematics.

The normal use of statistics is to indicate trends. The statistician finds (or rather selects) two or more phenomena which occur, or increase, more or less simultaneously. If these phenomena are to be dealt with statistically, the essential preliminary is to establish some causal connection between them. Without that any statistical treatment is sheer waste of time if not dishonest.

One example is the opinion that cigarette smoking and the incidence of lung cancer are connected. Let me say at once that I do not disagree with this opinion; it seems very probable that the two are connected. But no carcinogenic element in tobacco has as yet been isolated; the evidence, such as it is, is purely statistical. Over a certain period cigarette-smoking has increased; over the same period lung cancer has increasingly been diagnosed. *Therefore* cigarettes cause lung cancer. But over the same period there has been an enormous proliferation of television aerials. *Therefore* television causes lung cancer.

Certainly nobody is going to believe the latter assertion, because it would be difficult if not impossible even to conceive of a causal connection between television and lung cancer, whereas a correlation between smoking and a lung condition does not affront our common sense. But this is the fallacy of *post hoc ergo propter hoc*. We are arriving at our conclusion by assuming it in advance.

I find it difficult to understand, apart from the statistics themselves, the conclusions which doctors and other scientists draw from them. The result of an experiment conducted some years ago showed that of a colony of mice subjected for a month to a nicotine concentration equivalent to 20,000 cigarettes, half died of cancer. From this we deduce – what? Presumably that if you smoke fewer than 20,000 cigarettes a month you have an even chance of escaping lung cancer.

Statistics when designed to show that A causes B can mislead you in at least three ways. We will take it that A and B do tend to occur (or to increase) simultaneously.

Of course this may well be because A causes B. But may it not equally well be because B causes A, or because both A and B are effects of a perhaps unsuspected cause C? Or may it not (as in the case of cancer and television) be purely coincidental?

While I cannot cite any statistical evidence I am reasonably certain that the incidence of obesity is higher among people who wear gold watches than among the population taken as a whole, whereas their children are less liable than those of the majority to suffer from rickets. Does this prove that gold watches are bad for the figure while constituting a sovereign specific against rickets? Not at all. The possession of a gold watch is an indication of some degree of affluence, and the affluent are more likely than the impoverished to overeat. Rickets is a vitamin deficiency disease and hence practically unknown among the wealthy. The causal connection is only second-hand; except as a connecting link the gold watches can be ignored.

A recent article by a well-reputed doctor started off with the quite astonishing statement that there have been 'several estimates of addiction to alcohol or alcoholism

Obesity is a sign of affluence. Affluent people are more likely to possess gold watches than poor people; they are also more likely to overeat. There is, however, no causal connection to suggest that gold watches are bad for the figure.

(in Britain), ranging from 35,000 to 350,000'. What conceivable deductions can be drawn from such slipshod premises as these? The article goes on to tell us that of 300,000 alcoholics (what, one wonders, has become of the 35,000 estimate?) '70,000 showed evidence of physical or mental deterioration'. So, to quote the Duchess in *Alice in Wonderland* (a highly appropriate source), 'the moral of that is', if you are an alcoholic you have a better than three-to-one chance of being quite all right. One wonders if that is *really* what he wanted to convey!

Be very careful, by the way, of people who talk blithely about averages (except, of course, cricket averages and goal averages; they're all right). The word is so vague that to the mathematician it has several meanings while in the mouth of the uninstructed it is practically meaningless.

You have doubtless read from time to time in your newspaper references to the 'average' national income. This wholly illusory figure is, I presume, arrived at by adding together all our incomes, dividing by the number of people in the country, and thus coming up with a very simple answer which, like a good many other simple answers, is entirely valueless.

To show how misleading this sort of 'average' may be, let us take a hypothetical community of 10,000 people of whom 500 have incomes of £10,000 a year, 1,000 have £2,000 a year, 3,000 get £500 a year, 5,000 receive £350

a year, and the remaining 500 struggle along with a yearly pittance of £250. Let us suppose moreover that the minimum yearly income on which one can support life with any kind of decency is £400. A man who is either ignorant or dishonest will cheerfully and unhesitatingly tell you what is this community's 'average' income. Five hundred

GOAL AVERAGES
LEAGUE DIVISION II (Season 1967-8)

			Home		Goals	
Champions	Played	Won	Drawn	Lost	For	Again
Ipswich	42	12	7	2	45	20
Runners-up						
Queen's Pk. Rangers	**42**	**18**	**2**	**1**	**45**	**9**
Blackpool	**42**	**12**	**6**	**3**	**33**	**16**

people between them get £5,000,000; 1,000 receive £2,000,000; 3,000 get £1,500,000; 5,000, £175,000 and 500, £125,000, making a total of £10,375,000 for 10,000 people – an 'average' income of £1,037 10s a head; so they are quite comfortable. One does not have to be a left-wing socialist to perceive the absurdity of this conclusion. The

Won	Away Drawn	Lost	For	Goals Against	Points Total	Goal Average
10	8	3	34	24	59	1.795
7	**6**	**8**	**22**	**27**	**58**	**1.861**
12	**4**	**5**	**38**	**27**	**58**	**1.651**

Although the term 'average' is frequently misapplied, goal averages, like those shown above, do represent a true 'average'. For the benefit of non-footballing readers, the goal average is calculated by dividing the total of goals 'for' by the total of goals 'against'.

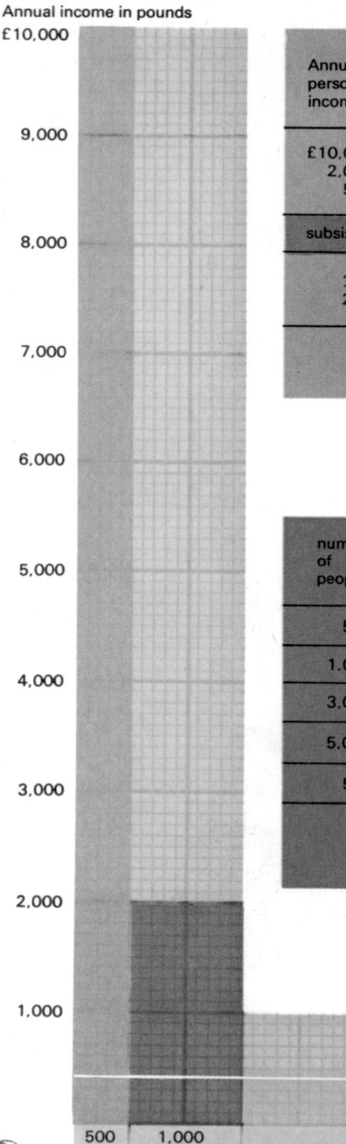

Table A

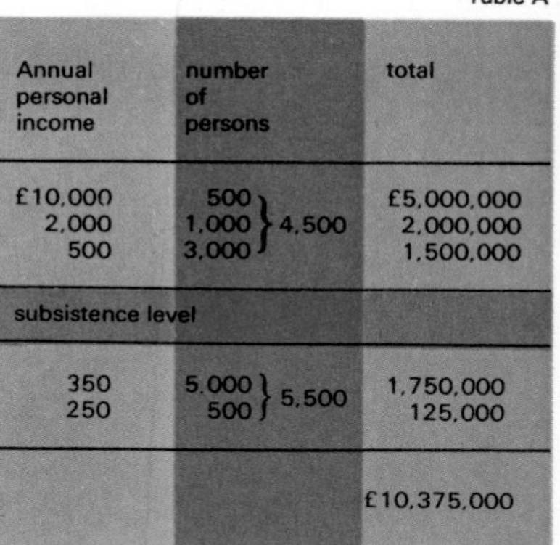

Annual personal income	number of persons		total
£10,000	500	4,500	£5,000,000
2,000	1,000		2,000,000
500	3,000		1,500,000
subsistence level			
350	5,000	5,500	1,750,000
250	500		125,000
			£10,375,000

Table B

number of people	percentage of total income	individual percentage
500	48·19	0·0964
1,000	19·28	0·0193
3,000	14·46	0·0048
5,000	16·87	0·0036
500	1·20	0·0024
	100%	

fact is that of these 10,000 people more than half are getting less than £400. If that is prosperity we can do without it; and if mathematics can lead us to no more sensible a conclusion than this, perhaps we might just as well do without mathematics too.

To the mathematician an 'average' may be a *mean* or a *median*; and if it is a mean it can be an arithmetic, a harmonic or a geometric mean. We shall come back to these terms in a moment; suffice it to say here that the mathematician selects the sort of average he will use according to the sort of problem which he is tackling at that particular moment.

In the case of our hypothetical community the mathematician will take as his 'average' the median, that is to say, the most prevalent figure (£350 a year). This produces the table shown in the upper panel opposite. From this it is clear that more than half the people are living sub-standard lives. Or our investigator might tabulate the facts in another way, see the lower panel. Perhaps clearer still is the subjoined graph, which needs only a glance to convince us that the economic health of this community falls far short of being satisfactory.

Remember that averages can be applicable only to phenomena with something in common, and only to their common element. Girls and boys are nowadays born in more or less equal numbers; if I am expecting an addition to my family can I therefore say that 'on the average' the child is likely to be bi-sexual? Again, dietitians have worked out for the normal healthy adult an optimum intake of food. If I eat like a pig and my wife starves, can I say that 'on an average' we as a couple enjoy a balanced diet?

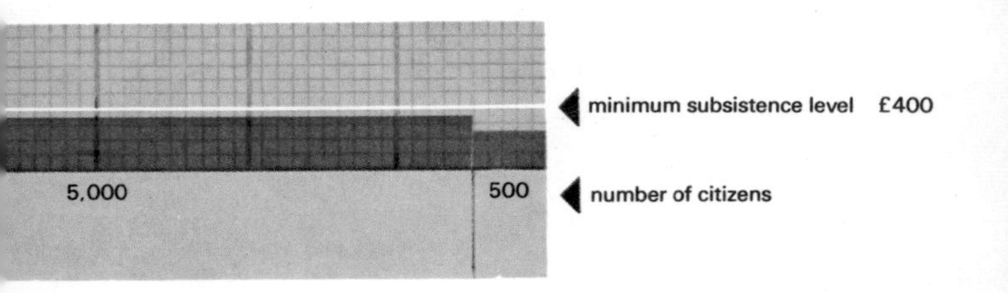

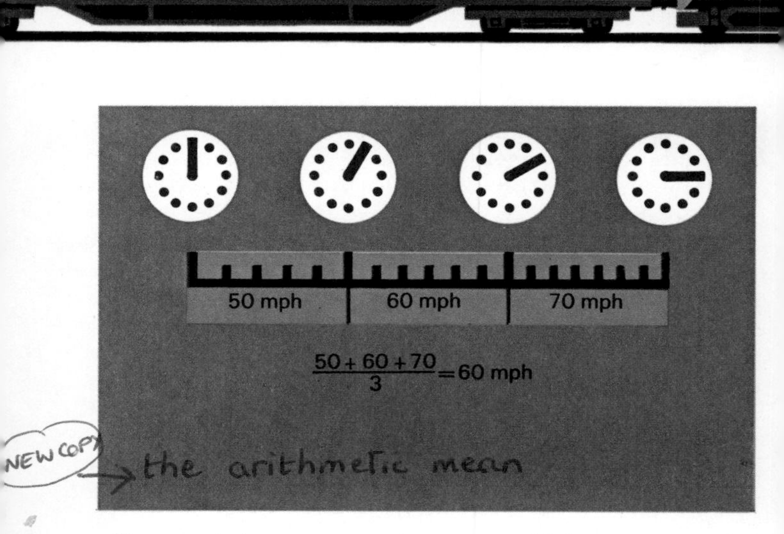

To revert, however, to the mean, which can be applied in some of the cases where 'averaging' makes sense. A train travels at 50 miles per hour during the first hour of its journey, at 60 m.p.h. during the second hour, and at 70 m.p.h. during the third. What is its average speed? Clearly the answer is 60 m.p.h. – the arithmetic mean, derived by adding the three distances and dividing by three:

$$\frac{50 + 60 + 70}{3} = 60.$$

But now take the proposition that the train travels at 50 m.p.h. for 300 miles and at 60 m.p.h. for the next 300 miles. Will its average speed be $\frac{50 + 60}{2} = 55$ m.p.h.? No. Because at 50 m.p.h. it will cover the first 300 miles in 6 hours and at 60 m.p.h. the second 300 miles will take 5 hours, so that it will travel 600 miles in 11 hours, with an average speed of $\frac{600}{11} = 55\frac{5}{11}$ m.p.h. The difference is accounted for by the fact that, although covering the

FAULTY WORKING ON THESE PAGES HAS ALREADY BEEN CORRECTED.

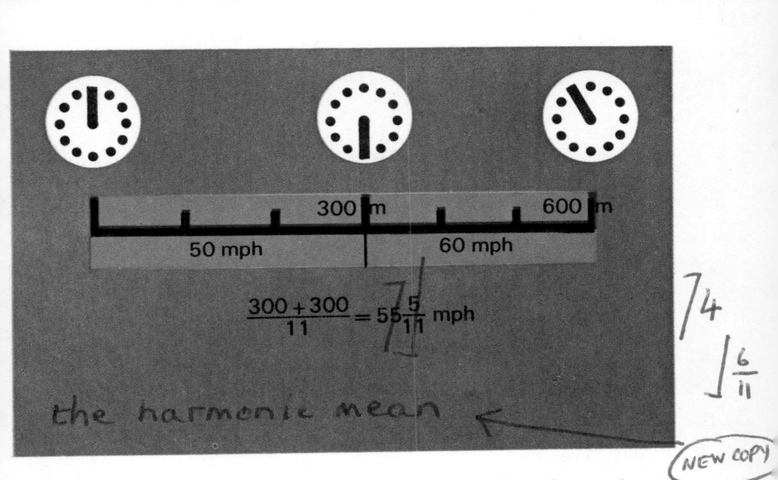

same *distance*, the train has been travelling for a longer *period* at the lower speed.

The answer is derived from what is called the harmonic mean: we have added together not the respective rates of speed but their reciprocals.* The harmonic mean is the number of values (2 × 300) divided by the sum of the two speeds, *i.e.*, $\frac{600}{110} = 55\frac{5}{11}$.

The geometric mean comes into play when we are considering questions of growth.** Supposing a town with a population of 10,000 in 1967 had increased its population to 13,000 by 1969. What is its population likely to have been in 1968?

Here neither the arithmetic mean nor the harmonic mean will help us. It is true that the population has

* Refer to the section on 'Sets and Series' for arithmetical, geometrical and harmonic progressions.

** See section on 'How Things Grow'.

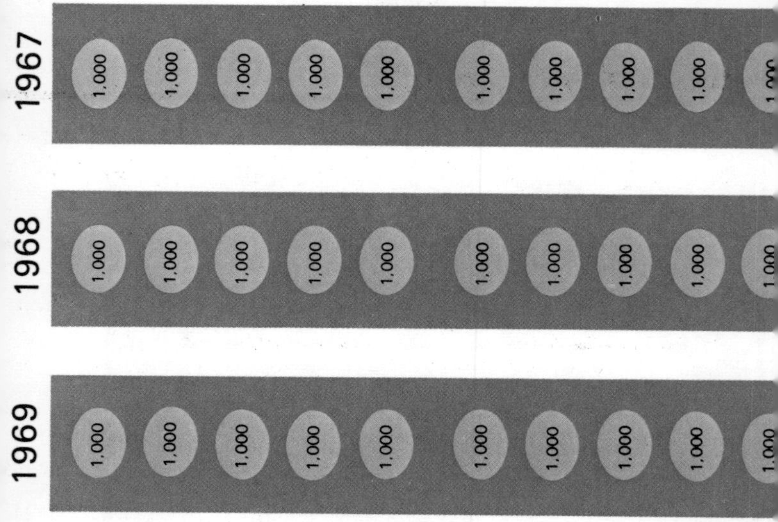

increased by 3,000 in two years, and if you use the arithmetic mean you may suppose the increase to have been at the rate of 1,500 a year. But this is clearly most unlikely. Younger people begin to reproduce, and populations (think of the 'population explosion') increase at compound interest.

To answer our question we may take it to begin with that the population of our town increases at the rate of x per cent per year. So, having started at 10,000 in 1967, it will have reached $10{,}000 \times \dfrac{100 + x}{100}$ in 1968, and therefore to $10{,}000 \left\{\dfrac{100 + x}{100}\right\}^2$ in 1969.

Thus: $13{,}000 = \dfrac{10{,}000(100 + x)^2}{10{,}000} = (100 + x)^2.$

This gives us the quadratic equation

$$x^2 + 200x - 3{,}000 = 0,$$

giving 14 as the approximate value of x. Thus the population in 1968 was approximately $\dfrac{10{,}000 \times 114}{100} = 11{,}400.$

In dealing with statistics there is another difficulty with which we have to contend, namely the fact that all of us, being human, are liable to error; and here I am referring to practical as distinct from theoretical error. If we have to measure a large number of quantities, whether linear or otherwise, we shall almost unquestionably make a mistake or two. We cannot prevent this; it is inevitable. But we can make allowances for it and thus arrive at a result that will take into account our human failings.

The greater the number of measurements you take the greater will be the probability that you have made mistakes; but the greater also will be the probability that you have made the same number of mistakes plus as minus so that they cancel out and your result, despite faulty working, may yet be correct. This applies in many branches of statistical work, for example, sampling, and the incidence of shots on a target, with many bunched around the middle and relatively few in the outer circles.

Suppose, for example, you have to count five pounds' worth of pennies into 100 piles of twelve each, and suppose furthermore that there is a one-in-ten chance that you will make a mistake of a penny (not more) in any one pile. Then it is reasonable to suggest that you are as likely as not to have made a mistake in ten out of the hundred piles; and of course you are equally likely to have in any individual case put in a penny too much as a penny too little. What is the probability that you will nevertheless have counted your five pounds accurately?

Of your ten piles possibly miscounted it may be that every one is a penny short, causing a total shortage of tenpence; that is likely to happen in one out of all possible distributions. Then again, you may be one short in each of nine piles with one over in the remaining one, accounting for a net shortage of eightpence; since your 'plus' pile may be any one of the ten, there are ten ways in which this can happen. Now if you are short in eight out of the ten piles and over in the remaining two, being sixpence short in all, this can happen in $\frac{10.9}{2} = 45$ ways.

If you carry on counting the possibilities in this way you will find that the successive numbers correspond with the coefficients of the terms in a binomial expansion (see Appendix II). They will, in the case we are now considering, be 1, 10, 45, 120, 210, 252, 210, 120, 45, 10 and 1, a total of 1,024 cases, among which 252 will be correct. So that your probability of arriving at a correct result is $\frac{252}{1024} = \frac{63}{256}$, or approximately three to one against. On the other hand, your probability of having an error not exceeding 2 is $\frac{210 + 252 + 210}{1024} = \frac{672}{1024} = \frac{21}{32}$; practically two to one on.

This is illustrated graphically in what is called the normal curve of error, which in the case we are now discussing is shown in the diagram opposite.

The curve may, of course, be steeper or flatter than that shown here, according to the data arising from the particular problem with which you are dealing; but it will always be a member of the 'family' of the normal curve of error, whatever the shape of the curve may be.

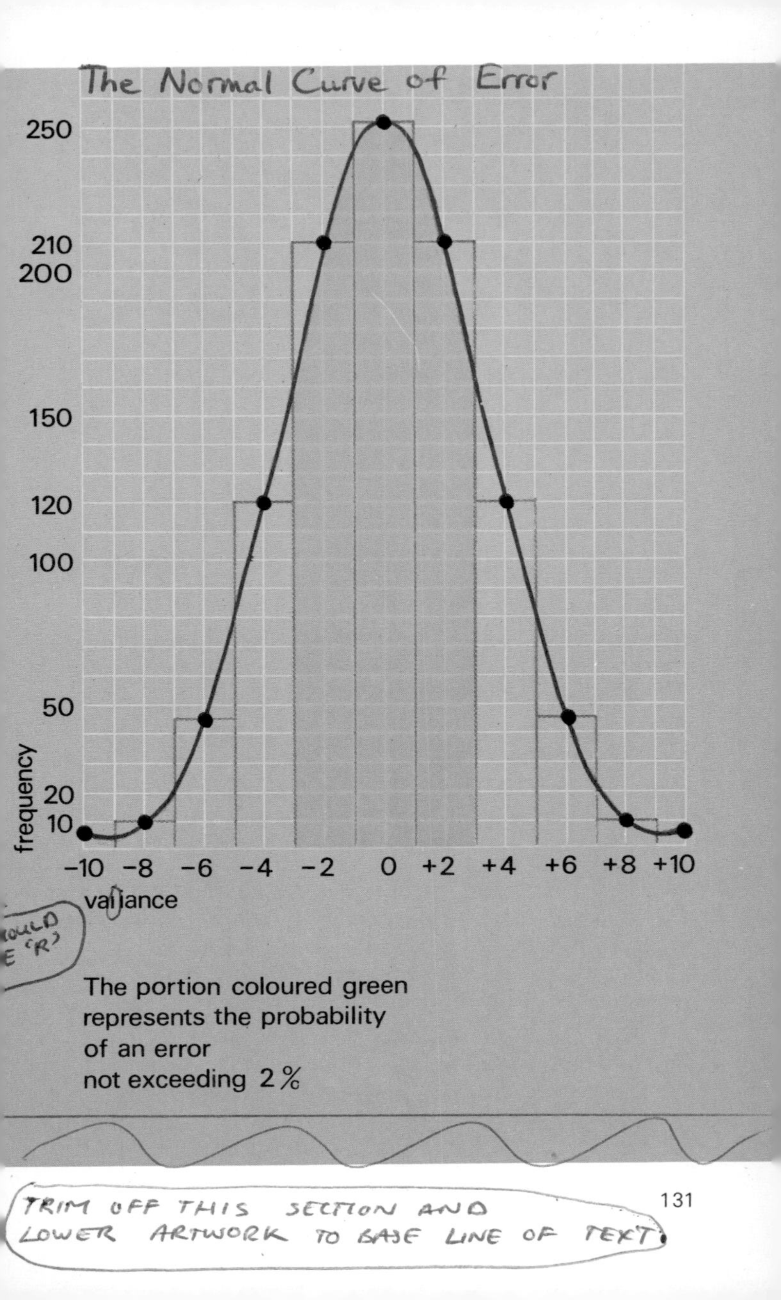

The portion coloured green represents the probability of an error not exceeding 2 %

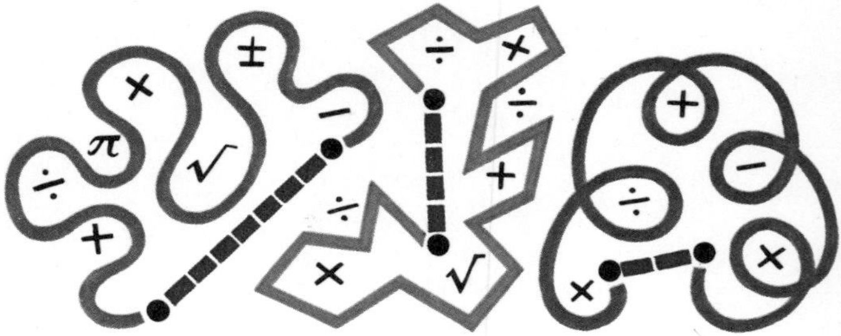

SOME SHORT CUTS

'Whatever you teach, be brief' – *Horace*

To multiply by 5, add one 0 and divide by 2.
To multiply by 25, add two 0's and divide by 4.
To multiply by 125, add three 0's and divide by 8, etc.
Example: 3,087 × 125 = 3,087,000 ÷ 8 = 385,875.
In the same way,
To divide by 5, put in one decimal point and multiply by 2.
To divide by 25, put in two decimal points and multiply by 4.
To divide by 125, put in three decimal points and multiply by 8, etc.
Example: 98,425 ÷ 25 = 984·25 × 4 = 3,937.

Factorizing

We consider for the most part only what *primes* (numbers divisible only by unity and themselves) are exact divisors of a given number; if we have established, for example, that a number is not divisible by 3, then we do not have to worry about the possibility of 6 being a factor.

A number is divisible by 2 only if its last digit is even.

A number is divisible by 3 only if its digital root* is 3 or a multiple of 3.

* The digital root of a number is arrived at by taking the sum of all its digits, then (if that sum consists of more than one digit) adding again, and so on until the result is a single figure. Thus, the digital root of 51,837 is 5 + 1 + 8 + 3 + 7 = 24; 2 + 4 = 6.

A number is divisible by 5 only if its last digit is 5 or 0.

A number is divisible by 9 only if its digital root is 9.

In fact this leaves us with only one number less than 10 for which there is no good rule of divisibility and that is 7.

A number is divisible by 11 only if the totals of the two sets of alternate digits are equal or differ by 11 or a multiple of 11 (*e.g.*, ABCDE is divisible by 11 only if A + C+ E− (B + D) = 0 or $11k$*).

The tests for divisibility by 3, 9 and 11 may need some explanation. Any number can be written as $a + 10b + 100c + \ldots$, a, b, c, etc. having any integral values you please (including 0) – *e.g.*, $1{,}057 = 7 + (5 \times 10) + (0 \times 100) + (1{,}000 \times 1)$. Now, if $a + b + c + \ldots = 3k$, we have:

$$a + 10b + 100c + \ldots = n$$
$$9b + 99c + \ldots = 9p.$$

Subtracting, $a + b + c + \ldots 3k = n - 9p$

so that $n = 9p - 3k.$

Divisibility by 9 is proved in the same way. To test for divisibility by 11,

$$a + 10b + 100c + 1{,}000d + \ldots = n$$
$$- a + b - c + d - \ldots = 11k.$$

Adding, $11b + 99c + 1{,}001d + \ldots = 11p$

so that clearly n (which is equal to $11k + 11p$) is divisible by 11.

A number is never a square unless it ends in 00, 1, 4, 25, 6 or 9 and its digital root is 1, 4, 7 or 9. If it ends in 4 or 9 the preceding digit must be even, if in 6 it must be odd. A number is never a cube unless its digital root is 1, 8 or 9.

Finally, a little curiosity, although it is not particularly useful because with numbers of more than three digits it becomes unwieldy. To square any number ending in 5, write down 25 and on the left of it the other number multiplied by a number one greater than itself, *e.g.*, $85^2 = (8 \times 9) \ldots 25 = 7{,}225$.

Of course, if you are concerned only with approximate results the best short cut is a slide-rule, a device we discussed in some detail on pages 40 and 41.

*k is the symbol for a constant.

A DIVERSITY OF DIVERSIONS

'Variety's the very spice of life' – William Cowper

Mathematics is by no means a finished art (or science) and there are many problems that still confront the amateur. I do not include here such old favourites as 'squaring the circle' (that is, by Euclidean methods, using only a straight-edge and compasses). Not only has this feat never been accomplished but it has been proved to be impossible – although it can be achieved with a more sophisticated armoury. There are, however, certain tasks that have so far defeated the mathematician but the impossibility of which has never been demonstrated.

Most famous of these is what is known as 'Fermat's Last Theorem'. Most of us are familiar with the equation of the right-angled triangle: $x^2 + y^2 = z^2$. Fermat noted in the margin of a book he was reading that he had devised a proof that there could be no rational solution to the equation $x^n + y^n = z^n$ if n were greater than 2. Unhappily the margin was not wide enough for the proof to be written in it; and for the past 300 years mathematicians have toiled in vain either to prove or to disprove Fermat's theorem, although it has been demonstrated for quite a number of values of n: it is known, for example, that there are no rational solutions for the equations $x^3 + y^3 = z^3$ or $x^4 + y^4 = z^4$. But the general theorem is still open to any student who cares to attempt it. Frankly, I would advise you to leave it alone.

Another unsolved problem relates to 'perfect' numbers. These are numbers that are equal to the sum of all their factors, including unity but not including the numbers themselves: for example, the factors of 6 are 1, 2 and 3, and $1 + 2 + 3 = 6$. The next 'perfect' number is 28, which again is equal to the sum of its factors: $1 + 2 + 4 + 7 + 14 = 28$.

So far as is known, all 'perfect' numbers are of the form $2^{p-1} (2^p - 1)$ where p is a prime number. We see this in the case of the two examples given as above: $6 = 2^{2-1} (2^2 - 1) = 2.3$; $28 = 2^{3-1} (2^3 - 1) = 4.7$.

The next 'perfect' number is $496 = 2^{5-1}\ (2^5 - 1) = 16.31$, of which the factors are 1, 2, 4, 8, 16, 31, 62, 124, 248, all of which add up to 496. But nobody has ever been able to prove that *all* 'perfect' numbers are of this kind: there may (or may not) be odd ones among them. This is another field for you to explore.

Here are a number of other problems that are by no means insoluble; in fact all of them can be dealt with by using no mathematical techniques other than those that have already been explained in this book. I hope that in addition to affording you some amusement they may help to illustrate some of the points I have raised. The answers will be found on page 145.

1 Garden plot

Here is a plan (not drawn to scale) of an irregular piece of land I am proposing to acquire. The only details regarding it that I have been able to obtain are that the angles at B and D are right angles and that each side measures an integral number of yards.

What is its least possible area? And can you construct another integral irregular quadrilateral of twice that area (again, of course, with diagonally opposite right angles just as in the first quadrilateral)?

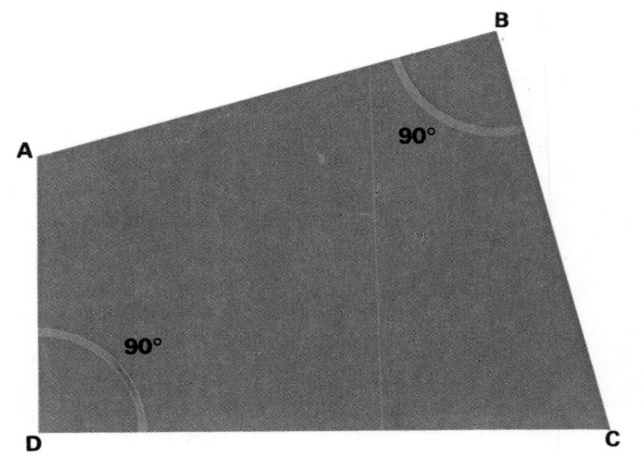

2 Sympathy and Antipathy

'Sympathy and Antipathy' is a gambling game of a very simple nature. One player bets an even stake on Sympathy or Antipathy and the other player turns up two cards from the pack. If they are both red or both black, Sympathy wins; if they are of different colours, then Antipathy wins.

My friend Professor Summit at once realized that Antipathy is the better bet, since whatever the top card of the pack may be there will remain only 25 of that colour as against 26 of the other. Accordingly he sat down to play, betting a shilling each time on Antipathy.

I returned to the room a few minutes later and asked how he was getting on. 'Badly', said the Professor. 'I've lost every hand so far.'

'Are you going to play through the pack on your system?' I inquired.

'I hardly know what to do,' the Professor told me,

‘because on the next turn-up Sympathy and Antipathy are even chances.’

How much had the Professor lost?

3 The infernal triangle

How many triangles, whatever their size or position, are there in the diagram below?

It is easy enough, you may say, to count them. So it is, though you may quite possibly find you have overlooked one.

But suppose it were a very large triangle, with a base of, say, 20 units. Counting them would then be an extremely arduous procedure. Can you find a formula to give the total number of triangles in an enclosing equilateral triangle of *any* size?

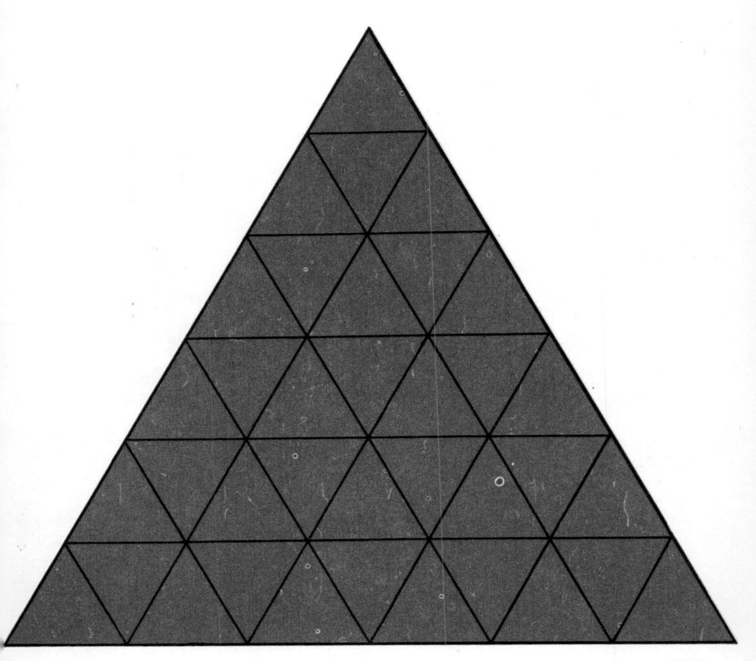

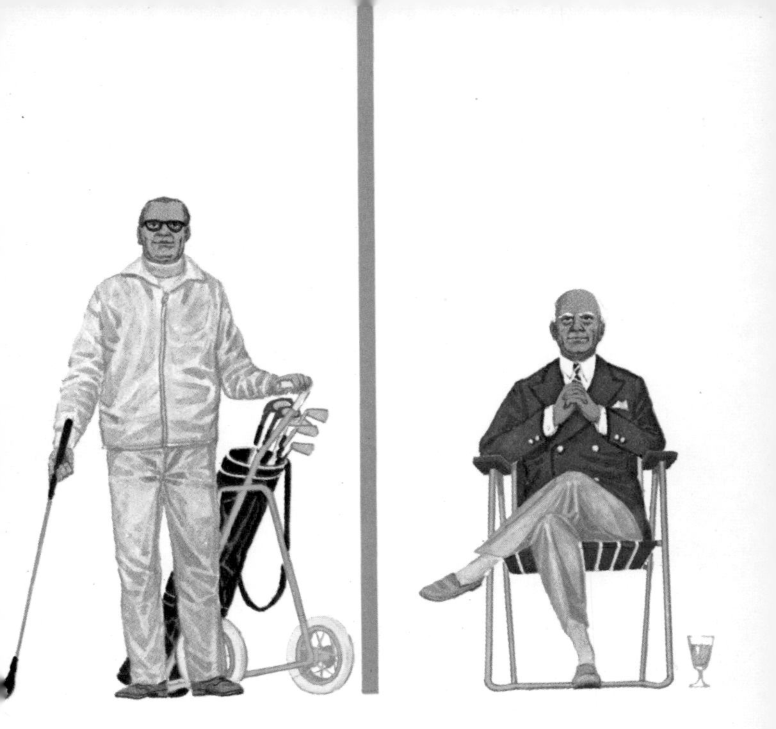

4 Sporting statistics

Propping up the bar of our local one day, the secretary of the Mudford Amateur Athletic Association said to me, 'Of the hundred members of our club, 90 play golf, 80 cricket and 70 tennis.'

'Very creditable,' I remarked, suppressing a yawn. 'Though I suppose, like most clubs of that kind, you have some inactive members?'

'Very few,' he answered. 'In fact there are 19 times as many members who play all three games as there are of those who do not play at all. All those who confine themselves to one game are golfers.'

How many of the members of the Mudford Amateur Athletic Association play three games, two games, and one game, and what games do they play?

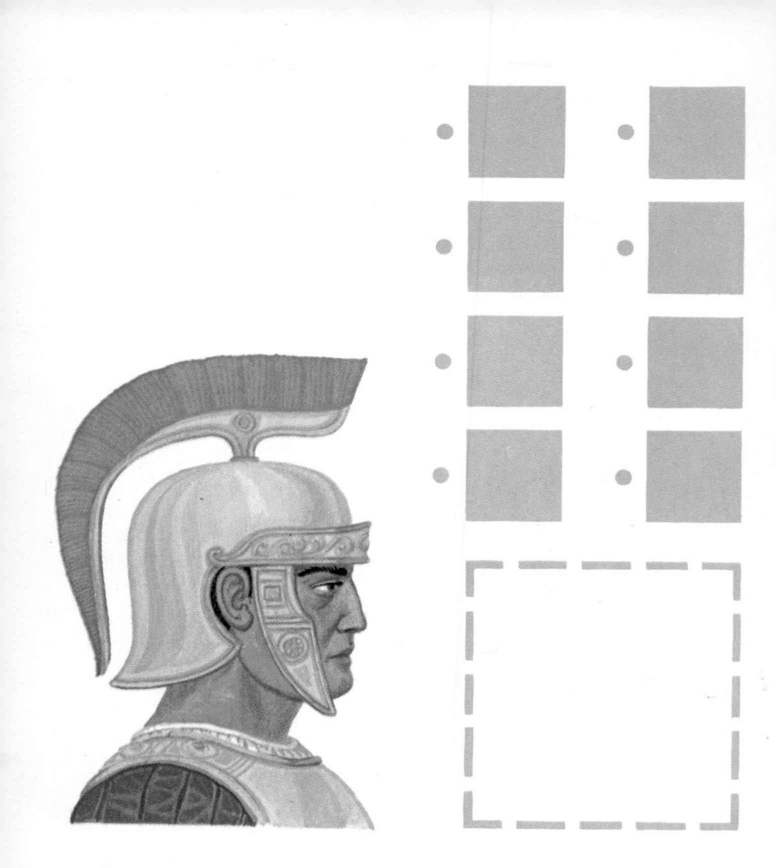

5 Democracy in the ranks

The troops of the great Roman general Pomponius Mathematicus paraded for his inspection in eight solid squares, each led by an officer. But when Pomponius arrived he insisted on a more democratic procedure. He instructed his junior officers to fall in with their men, an example which he himself followed. After that the entire army marched off in one solid square together.

Bearing in mind the fact that a Roman general would certainly not have commanded fewer than a hundred men, what was the smallest muster of troops that could have carried out this manoeuvre?

6 Grandpa's little wager

'Two of your grandchildren are coming to my party tomorrow,' I told Prodger.

'Are they?' asked the old man, who is always ready for a bet. 'I'll lay you two to one they won't both be girls.'

'I'm not a betting man,' I told him.

'Look here,' the old chap went on eagerly, 'I'll lay you five to one they won't both be boys.'

Assuming that Prodger laid the correct odds in both cases, how many grandsons has he?

7 Down on the range

The members of our rifle club make up in keenness whatever they may lack in skill.

In the semi-finals, which were competed for at the Blunderbuss Range, it was agreed that the points awarded should be seven for a bull, five for an inner, three for a magpie, and one for an outer.

None of the four competitors, I am happy to say, missed the target with any shot. Alfred did not once hit the bull, but registered hits in all the other circles; Bertram scored everything except an inner; Charles's score included everything except magpies, and David's shots missed only the outer ring. Each, of course, fired the same number of rounds, and they all finished up with the identical score – 30 points.

How many bulls were scored by the man who had three magpies?

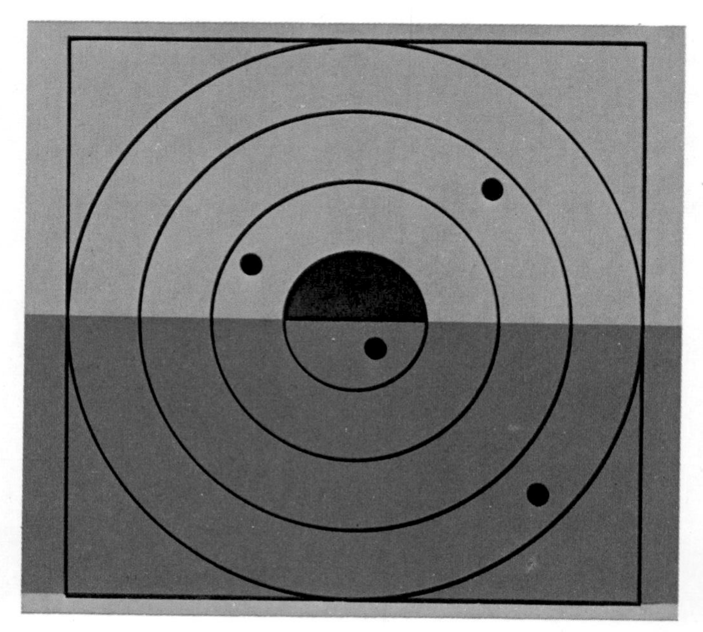

8 What price oranges?

'Your oranges are expensive,' I remarked to my greengrocer yesterday.

'I know they are, sir,' he replied. 'They're fetching high prices in the market. Now, if my wholesaler would let me have 16 more oranges for a pound I could drop my charges by $1\frac{1}{2}$d an orange and still make the same rate of profit.'

Today I was pleased to see that he had lowered his prices by that amount. However, I still didn't think them cheap, and told him so.

'True enough, sir,' he answered. 'When I can get another 16 for a pound I shall be able to knock one penny more off my prices without affecting my rate of profit.'

How much is my greengrocer currently charging for oranges, and what rate of profit does he make?

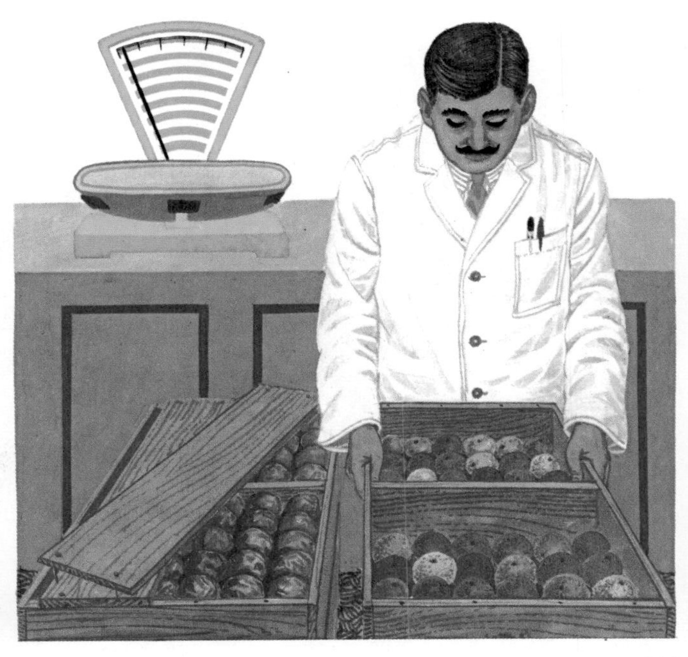

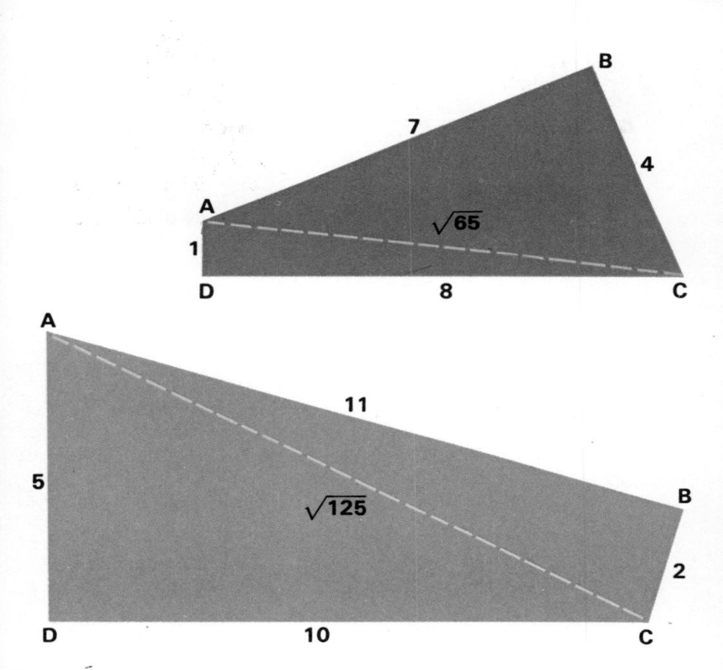

Solutions

1 Garden plot

The plot of land offered to me has an area of 18 square yards, with sides measuring 1, 4, 7 and 8 yards respectively. A similar irregular quadrilateral of area 36 square yards would have sides of 2, 5, 10 and 11 yards respectively.

The first problem resolves itself into finding the lowest number that is the sum of two integral squares in two ways. This is 65, ($1^2 + 8^2$ or $4^2 + 7^2$). The resultant quadrilateral consists of two triangles with base 8 yards and altitude 1 yard and base 7 yards and altitude 4 yards respectively: $\frac{8 \times 1}{2} + \frac{7 \times 4}{2} = 18$.

Equally $125 = 10^2 + 5^2$ or $11^2 + 2^2$, giving two triangles of areas 25 and 11 square yards respectively, a total of 36 square yards.

2 Sympathy and Antipathy

Suppose there remain in the pack x red cards and y black ones. Then the probability of turning up two red cards is $\frac{x(x-1)}{(x+y)(x+y-1)}$ and the

probability of drawing two black cards is $\frac{y(y-1)}{(x+y)(x+y-1)}$. Thus the probability of turning up Sympathy is the sum of these two, or $\frac{x(x-1)+y(y-1)}{(x+y)(x+y-1)}$. In the same way the probability of turning up two cards of different colours (Antipathy) is $\frac{2xy}{(x+y)(x+y-1)}$.

Hence we have $x^2 - x + y^2 - y = 2xy$, or $(x-y)^2 = x + y$. But we know the number of cards left in the pack to be even (because two are turned up each time). Thus $x + y$ is an even square less than 52. It must therefore be 36, 16 or 4.

If	$x + y =$	36	16	4
	$x - y =$	6	4	2.
So	$x =$	21	10	3
	$y =$	15	6	1.

But x and y are both even (since Sympathy has won every time), so the only possible answer is that there remain 10 (or 6) red cards and 6 (or 10) black ones. The Professor has lost 18 successive coups and 18 shillings.

3 The infernal triangle

The total number of all triangles in one with a base of 6 units is 78; in a 20-unit size triangle there are 2,255.

Formulae for triangles of *any* size (taking n as the number of small triangles with their bases on the base line) are the following:
when n is even

$$S = \frac{2n^3 + 5n^2 + 2n}{8}.$$

when n is odd

$$S = \frac{2n^3 + 5n^2 + 2n - 1}{8}.$$

If you think it neater to have only one formula to cover all cases, then

$$S = \frac{4n^3 + 10n^2 + 4n - 1 + (-1)^n}{16}.$$

4 Sporting statistics

Of the 100 members 57 play golf, cricket and tennis, 16 play golf and cricket only, 6 golf and tennis only, 7 cricket and tennis only, 11 confine themselves to golf, and there are three non-playing members.

Even if the 10 non-golfers all play cricket there must still be 70 who take part in both games. If the 30 who do not participate in both golf and cricket all play tennis, there must still be at least 40 members who take part in all three games. Equally, the number cannot exceed 70 (the total number of tennis players). Thus we have to find a number not less than 40 and not greater than 70 which is exactly divisible by 19. Only one number, 57, will meet the case. Thus there are 57 members who play all three games and three who 'only stand and wait'.

Of the remaining 40 players some will play two games only and some only one. Let a be the number who play golf and cricket only, b the number who play golf and tennis only, c the number who play cricket and tennis only, and d the number who only play golf.

Then $a + b + c + d = 40$.

Total number of golfers $= 57 + a + b + d = 90$.

So $a + b + d = 33$.

Total number of cricketers $= 57 + a + c = 80$.

So $a + c = 23$.

Total number of tennis players $= 57 + b + c = 70$.

So $b + c = 13$.

These are ordinary simultaneous equations, giving $a = 16$, $b = 6$, $c = 7$, $d = 11$.

5 Democracy in the ranks

There were eight squares of 324 (18^2) men, which with Pomponius and the other eight officers made a grand total of 2,601 (52^2). Had it not been for the provision that Pomponius must have commanded at least a hundred men, the problem could have been solved by having eight squares of nine men each, which with their general and the eight junior officers would make a total of $(8 \times 9) + 9 = 81 = 9^2$.

The problem resolves itself into finding integral solutions to the equation $8x^2 + 9 = y^2$.

It is interesting to speculate whether there is any general formula when there are x squares with y^2 men in each. There is never a solution when $x = 4k + 1$ or $4k + 2$. There is always a solution or solutions when (and I suspect only when) $x = k^2 - 1$, $k^2 + 3$, or $4k$.

There is, of course, the mathematically correct but actually absurd solution $4 \times 1^2 + 5 = 9$. But while '1' is unquestionably a square number, we can hardly envisage 'squares' each consisting of one soldier.

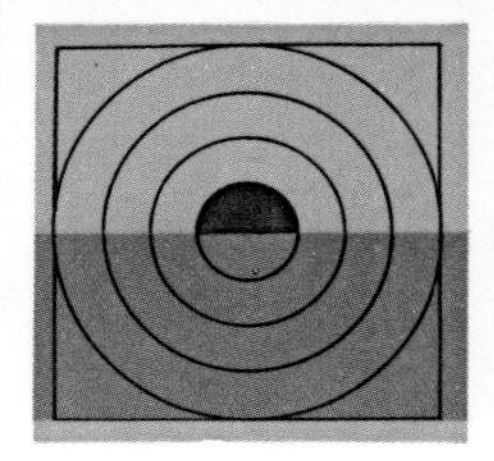

6 Grandpa's little wager

Prodger has 36 grandchildren, 21 of them girls and 15 boys.

Let the number of girls be represented by x and the number of boys by y. Then the total number of grandchildren will be $x + y$.

The probability of the first child I meet being a girl is $\frac{x}{x+y}$; the probability of the second being a girl is $\frac{x-1}{x+y-1}$. Thus the probability of their both being girls is the product of these two fractions, that is to say $\frac{x(x-1)}{(x+y)(x+y-1)} = \frac{1}{3}$. In the same way the probability of their both being boys is $\frac{y(y-1)}{(x+y)(x+y-1)} = \frac{1}{6}$.

From this we see that $3x(x-1) = 6y(y-1)$, and hence that $\frac{x(x-1)}{2} = y(y-1)$. Now $\frac{x(x-1)}{2}$ is a triangular number and $y(y-1)$ is both a triangular and half a triangular number. A value of 15 for y is appropriate, giving 21 for x: $\frac{15.14}{36.35} = \frac{1}{6}$, and $\frac{21.20}{36.35} = \frac{1}{3}$.

7 Down on the range

Clearly Alfred (no bulls) fired at least 8 shots and David (no outers) cannot have fired more than 8 shots. So everybody fired 8 times.

Suppose Alfred scored x inners, y magpies and z outers. Then

$$\begin{aligned} 5x + 3y + z &= 30 \\ \text{and} \quad x + y + z &= 8. \end{aligned}$$

Thus Alfred's score was either 5 inners, 1 magpie and 2 outers or 4 inners, 3 magpies and 1 outer.

Similarly, the possible scores of the other competitors are:

A		A	
5 inners (5 pts)	25	4 inners (5 pts)	20
1 magpie (3 pts)	3	3 magpies (3 pts)	9
2 outers (1 pt)	2	1 outer (1 pt)	1
8	30	8	30

B	
3 bulls (7 pts)	21
2 magpies (3 pts)	6
3 outers (1 pt)	3
8	30

B	
2 bulls (7 pts)	14
5 magpies (3 pts)	15
1 outer (1 pt)	1
8	30

D	
1 bull (7 pts)	7
1 inner (5 pts)	5
6 magpies (3 pts)	18
8	30

We need not concern ourselves with Charles, since we know he scored no magpies.

Hence the only man with three magpies was Alfred, who scored no bulls.

8 What price oranges?

Today my greengrocer is charging sixpence each for oranges he buys at 80 for a pound, his rate of profit being 100 per cent.

A short cut to the result is to assume that my greengrocer is not paying less than the fourth part of a penny for an orange. Then the number he gets for a pound must be an aliquot part of 960, and we must find three of these with a common difference of 16. There are in fact six of them: 16, 32, 48, 64, 80 and 96. Experiment shows that the three numbers 64, 80 and 96 fit the data.

For those who like a more rigorous method, let the number he bought for £1 on the first day be x and his rate of profit be y per cent. Then each orange cost him $\frac{240}{x}$ pence and he sold it for $\frac{240}{x} \times \frac{100 + y}{100}$ pence. On the second day he bought at $\frac{240}{x + 16}$ and sold at $\frac{240}{x + 16} \times \frac{100 + y}{100}$; and on the third day he would be able to sell at $\frac{240}{x + 32} \times \frac{100 + y}{100}$.

So
$$\frac{240(100 + y)}{100x} - \frac{240(100 + y)}{100(x + 16)} = \frac{3}{2}$$
and
$$\frac{240(100 + y)}{100(x + 16)} - \frac{240(100 + y)}{100(x + 32)} = 1.$$

The solution to the simultaneous equations gives $x = 64$, $y = 100$.

Appendix I
EXTRACTION OF ROOTS

'The insane root that takes the reason prisoner' – Shakespeare

The most easily comprehensible method of extracting the square root of any number is by means of repeated approximations.

Suppose we want to find the square root of two (which, though 'real', will not be rational). Clearly it will be more than one and less than two (the square root of four). For a first approximation we may call it $1 + x$, remembering that x will be a fraction less than one. So we take $(1 + x)^2 = 2$, and thus that $1 + 2x + x^2 = 2$, or $2x + x^2 = 1$. Since x is a fraction less than one, x^2 will be quite small, and we may *for the moment* ignore it (not the same thing as calling it 0 and writing it off;* we realize that however small it is we shall have to consider it later). We therefore take $1 + 2x = 2$, or $2x = 1$ and $x = \frac{1}{2}$, giving $1 + \frac{1}{2}$, or $1\frac{1}{2}$ as the square root of 2. $(1\frac{1}{2})^2 = 2\frac{1}{4}$, so our approximation is not very good; it is clearly a little too large.

Taking a new (unknown) value for x, we may say $\sqrt{2} = (1\frac{1}{2} - x)$, or $2 = (1\frac{1}{2} - x)^2 = 1\frac{1}{2}^2 - (2.1\frac{1}{2})x + x^2 = 2\frac{1}{4} - 3x + x^2$, giving $- 3x + x^2 = - \frac{1}{4}$. The quantity represented by x^2 will be even less than before, so again we ignore it, leaving ourselves with $- 3x = - \frac{1}{4}$, or $x = \frac{1}{12}$. So this next approximation is $\sqrt{2} = 1\frac{1}{2} - \frac{1}{12} = 1\frac{5}{12}$.

$$\{1\tfrac{5}{12}\}^2 = \{\tfrac{17}{12}\}^2 = \tfrac{289}{144} = 2\tfrac{1}{144}.$$

This is close enough for most purposes; if the reader desires an even more accurate result he need only continue with the same procedure. Take $\sqrt{2} = 1\frac{5}{12} - x$ (x being now very tiny indeed and x^2 even smaller) and carry on as long as you please: we chose this value for $\sqrt{2}$ because, of course, $1\frac{5}{12}$ had proved just too big for accuracy.

In extracting square roots there is a quicker method. Write down the number and mark off *pairs* of digits from the right, so that $\sqrt{237169}$ then becomes $\sqrt{23,71,69}$.

* Refer to 'All About Nothing', page 58.

Now (step *a*) take the highest number of which the square does not exceed the first digit (or pair of digits): in this case it will be 4 (16 is less than 23; 5^2 (25) is greater), write 16 under the 23, and insert the number 4 on the left above the bar. Then (step *b*) subtract, and bring down the next two figures from the original number. Now (*c*) put to the left of this new number *double* the number you wrote above the bar (8) and on the right of it the highest number (x) that will make ten times that number plus x not greater than your new number (771); in this case it will be 8. Multiply by the new number and subtract (*d*); write the new number (8) above the bar. Step (*e*), bring down two more digits from the original number and write on the left double the whole number above the bar, followed by the number by which you may multiply ten times this and not exceed the number you have just written as the result of step (*e*). Continue if necessary; here you already have the result: $\sqrt{237169} = 487$.

```
                 487
               √23,71,69
(a)              16
(b)            | 771
(c)      88    |
(d)            | 704
               | 6769
(e)      967   | 6769
```

Use this method for a number without an integral (whole number) square root: just add a decimal point and as many noughts as you choose to get a suitable approximation:

```
                225·692...
              √5,09,37·00,00,00, ...
               4
42            |109
              | 84
445           |2537
              |2225
4506           |31200
               |27036
45129           |416400
                |406161
451382           |1023900
                 | 902764
                   121136
```

Testing this for accuracy, $(225{\cdot}692)^2 = 50936{\cdot}88$. Near enough, wouldn't you say?

To extract a cube root we proceed on similar lines, although the working is rather more tedious. To find, for example, $\sqrt[3]{2}$. We know that $1^3 = 1$ and $2^3 = 8$, so that $\sqrt[3]{2}$ will be between 1 and 2 and considerably closer to 1. So take $(1 + x)^3 = 2$, giving $1 + 3x + 3x^2 + x^3 = 2$. Here we disregard (remember, only temporarily) x^3 and write $1 + 3x + 3x^2 = 2$, so that $3x^2 + 3x - 1 = 0$. By the quadratic formula (see page 30) $x = \frac{-3 \pm \sqrt{9 + 12}}{6} = \frac{\sqrt{21} - 3}{6} = \frac{4{\cdot}58 - 3}{6} = \frac{1{\cdot}58}{6} = {\cdot}26$; so $1 + x = 1{\cdot}26$. Try cubing $1{\cdot}26$. The result is $2{\cdot}0004$; close enough for all reasonable requirements. But here again you can if you wish take it further.

Appendix II
THE BINOMIAL THEOREM

The Binomial Theorem has been ascribed variously to Newton, Leibniz and the Persian poet Omar Khayyám. For its working it depends largely on a construction known as Pascal's Triangle.

Having thus cleared the decks, we may go on to say that the Binomial Theorem is a method of writing down the expansion of the binomial (expression consisting of two terms) $(a + b)^n$, or $a + b$ multiplied by itself n times. It is valid for any value of n, positive or negative, integral or fractional, and it turns out to be:

$$(a + b)^n = a^n + na^{n-1}b + \frac{n(n-1)}{2!}a^{n-2}b^2 + \frac{n(n-1)(n-2)}{3!}a^{n-3}b^3 + \ldots.$$

This may seem not very easy to remember, but there are a couple of useful mnemonics that can be employed.

The powers of a and b present no difficulty: the first term is a^n, the second $a^{n-1}b$, the third $a^{n-2}b^2$, and so on; the power of a decreases by one in each successive term while that of b increases by one – the two powers (of a and b) will always make a total of n. The coefficients (the numbers preceding a and b in each term) may seem more complicated; but there is an easy way of remembering them. We construct the numerical triangle which has become associated with the name of Pascal by the simple method illustrated on this page; each line begins and ends with unity, and the intervening numbers are in all cases the sums of the two numbers above them in the previous line:

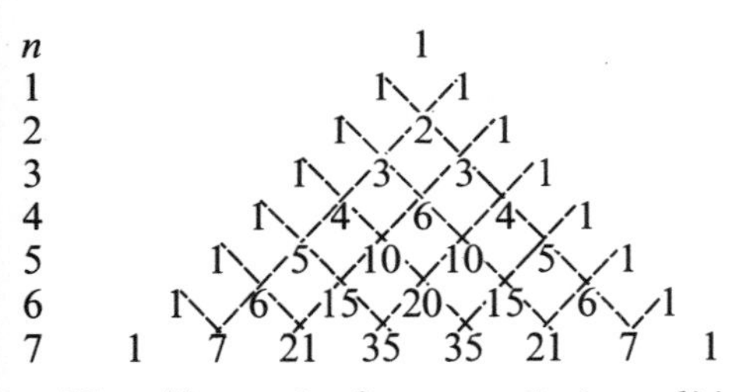

Now if we take for example $(a + b)^4$, we know that the five successive powers of a and b will be a^4, a^3b, a^2b^2, ab^2 and b^4, and the successive coefficients can be read along the relevant line ($n = 4$), thus being 1, 4, 6, 4 and 1. So the expansion of $(a + b)^4$ is

$$a^4 + 4a^3 b + 6a^2 b^2 + 4ab^3 + b^4.$$

There you have the essence of the Binomial Theorem. It is a little more difficult to operate with negative or fractional indices, but the same principles hold good (although Pascal's Triangle will not help here).

The Binomial Theorem can sometimes be used in ordinary arithmetical calculations. Suppose, for example, you want to find the value of $100\frac{1}{4}$ raised to the fourth power. Then, using line 4 of the Triangle, we find:

$$(100 + \cdot 25)^4 = 100^4 + 4.100^3(\cdot 25) + \ldots$$

and, without bothering to take the expansion any further, we get:

$$(100 + \cdot 25)^4 = 100{,}000{,}000 + 1{,}000{,}000 + \ldots$$
$$= 101{,}000{,}000 + \ldots$$

Worked out the 'hard way' the answer comes to 101,003,756 and a fraction, so that we have arrived at a good approximation (the error being less than one in 10,000) without much trouble.

GLOSSARY

Algebra. A method of calculation by symbols – generalized arithmetic.

Analytical Geometry (also *Co-ordinate Geometry*, or *Graphs*). The system of Descartes for indicating points on a plane with two co-ordinates, points in space with three.

Arc. Part of the circumference of a circle or other curve.

Arithmetical Progression. A series of numbers each differing from its predecessor by a constant difference (plus or minus).

Average. Meaningless word employed (and misused) by non-mathematicians. See *Mean*, *Median*.

Binary Scale. A system of numeration using only the numerals 1 and 0.

Binomial Theorem. A method of calculating the expansion of the term $(a + b)^n$.

Calculus. A system of calculation used in higher mathematics, discovered independently by Newton and Leibniz.

Characteristic. The integral part of a logarithm.

Circumference. The boundary line of a circle or other closed curve.

Combination. The number of ways of selecting (in any order) x objects from a total of n objects.

Complex Numbers. Numbers consisting of the imaginary number (i, or $\sqrt{-1}$) with a rational number.

Conic Sections. The plane figures which result from cutting a doubled cone in various directions.

Convergent Series. A series that as additional terms are added tends to approach a certain limit.

Co-ordinate Geometry. See *Analytical Geometry*.

Cosine. The trigonometrical ratio arrived at by dividing the base of a right-angled triangle by its hypotenuse.

Cubic Equation. An equation containing x to the power of 3 but no higher power.

Cylinder. A solid figure generated by a straight line remaining parallel to a fixed axis and moving round a closed curve.

Decimal Notation. Numeration in sets of ten.

Digital Root. The figure arrived at by adding all the digits of a number, then (if the result exceeds 9) adding again, and so on.

e. The exponential number: the sum of the series $1 + \frac{1}{1} + \frac{1}{1.2} + \frac{1}{1.2.3} + \ldots$ tending to the limiting value of 2.71828 ...

Ellipse. An oval-shaped conic section.

Equilateral Triangle. A triangle having three equal sides (and angles).

Figurate Numbers. Numbers which if depicted by a dot for each unit form regular plane figures (triangles, squares, etc.).

Geometrical Progression. A series of numbers each of which stands in a constant ratio of multiplication or division to its predecessor.

Geometry. That branch of mathematics which treats of spatial magnitudes.

Graphs. See *Analytical Geometry.*

Great Circle. A line round the earth having its centre at the earth's centre.

Harmonic Progression. A series the members of which are the reciprocals of those in an Arithmetical Progression.

Hyperbola. A double curve formed by a cut going through both branches of a double cone. See *Conic Sections.*

Hypotenuse. The line opposite to the right angle in a right-angled triangle.

i. The imaginary number.

Imaginary Number. $\sqrt{-1}$ (the square root of minus one).

Indices. Small figures written to the right and above numbers showing the power to which they are to be raised.

Integers. Whole numbers; the natural numbers whether positive or negative.

Isosceles Triangle. A triangle in which two of the sides are equal.

Logarithms. The power to which a fixed number (the base) must be raised to equal a given number. Invented by Napier.

Mantissa. The decimal part of a logarithm.

Mean (arithmetic). The result obtained by dividing the sum of a series by the number of terms in it.

Mean (geometric). The *n*th root of the product of *n* members of a series.

Mean (harmonic). The reciprocal of an arithmetic mean.

Median. That approximation to an 'average' that is most nearly central between the extremes.

Napierian Logarithms. Logarithms to the base *e*.

Natural Numbers. All positive integers, 1, 2, 3 and so on.

Operatives. Dynamic numbers used to carry out operations.

Pascal's Triangle. A triangular arrangement of numbers showing the coefficients in a binomial expansion.

Parabola. A plane curve formed by a cut made parallel to the side of a cone.

Perimeter. The boundary line or lines of a plane figure.

Permutation. The number of possible arrangements of *x* objects selected from *n* objects.

Pi (π). The ratio formed by dividing the circumference of a circle by its diameter, equal to approximately 3·14159.

Plane Geometry. That part of geometry that deals with figures on a plane surface.

Power. The number of times by which a number is to be multiplied by itself.

Prime Numbers. Natural numbers that have no integral divisors other than themselves and one.

Quadratic Equations. Equations containing x to the power of two but no higher power.

Rational Numbers. Numbers that can be expressed by the use of vulgar fractions or of finite or recurring decimals.

'Real' Numbers. All numbers to which an approximate value may be attributed, including infinite decimals but excluding imaginary and complex numbers.

Reciprocal. That number which when multiplied by a given number will produce unity (*e.g.*, $3 \times \frac{1}{3} = 1$, so that $\frac{1}{3}$ is the reciprocal of 3).

Recurring Decimals. All decimal fractions in which a figure or set of figures recurs in the same order.

Right Angle. An angle of 90 degrees.

Series. Sets of numbers in which each member varies by a constant or progressive difference from its predecessor.

Sine. The trigonometrical ratio arrived at by dividing the upright in a right-angled triangle by the hypotenuse.

Square Number. The product of any number multiplied by itself. See *Figurate Numbers*.

Tangent. (i) A straight line that touches but does not cut a curve. (ii) The trigonometrical ratio produced by dividing the upright of a right-angled triangle by the base.

Triangular Numbers. See *Figurate Numbers*.

Trinary Scale. A system of numeration using only the numerals -1, 0 and $+1$.

Vanishing Triangles. A construction used in summing series.

INDEX

Page numbers in bold type refer to illustrations.

OTHER TITLES IN THE SERIES

Natural History

Bird Behaviour
Fishes of the World
Fossil Man
Life in the Sea
Natural History Collecting
Prehistoric Animals
Seashores
Snakes of the World
Wild Cats

Gardening

Garden Flowers
Garden Shrubs
Roses

General Information

Sailing Ships
Sea Fishing
Trains

Popular Science

Atomic Energy
Computers at Work
Electronics

Arts

Architecture
Jewellery
Porcelain

Domestic Animals and Pets

Budgerigars
Pets for Children